SOTHIS PRESS

Sothis Press aims to completely reimagine the public domain book with an emphasis on universal access to captivatingly designed print editions. We believe that technology will enable every single book ever published to be perpetually accessible in beautiful print form. Sothis is beauty in ubiquity.

www.sothis.press
1 Waterhouse Square, London, EC1N 2ST

The Northern Pacific Railway, Its Effect Upon the Public Credit, the Public Revenues, and the Public Debt

Speech of Hon. William Windom, of Minnesota, Delivered in the House of Representatives, January 5, 1869

William Windom

SOTHIS PRESS

ISBN 978-1-0365-0123-5

The Northern Pacific Railway, Its Effect Upon the Public Credit, the Public Revenues, and the Public Debt
is in the public domain.

THE NORTHERN PACIFIC RAILROAD.

SPEECH

OF

HON. WILLIAM WINDOM,

OF MINNESOTA.

The House being in Committee of the Whole on the President's Message—

Mr. WINDOM said:

Mr. Chairman: I rise to advocate the claims of a great national enterprise, in which the people of my State and of the whole country have a profound interest; an enterprise which I will endeavor to show is in all respects practicable and meritorious; one that will cost the Government no money, but will save vast sums to the treasury, and which will strengthen the public credit, largely augment the public revenues, and reduce the public burdens. My time is limited and my facts are numerous; I will, therefore, proceed at once to its discussion. When Mr. Edwin F. Johnson, Governor Stevens, and others, first presented the claims of the Northern Pacific railroad, they encountered many objections and obstacles interposed by those who were not well informed in regard to it. Since then, however, the discussions here, and the surveys and explorations of the country through which the road is to pass, have thrown a flood of light upon the proposition, and to-day there is no great enterprise which lies nearer to the hearts of the people than this.

In attempting to point out some of the peculiar advantages of this route, I will not be understood as depreciating the merits of others, for I fully admit their inestimable value to the country, and will go with him who goes furthest in conceding to them every necessary and proper encouragement.

The Central line, born in the darkness and agony of the nation's struggle, and matured with an energy unprecedented, will ever stand one of the proudest monuments of American courage, enterprise, and statesmanship.

The Kansas Pacific road, yet in its infancy, with only some four hundred miles completed, has, in its beneficial results to the Government, far surpassed the hopes of its most sanguine friends, and fully demonstrated the wisdom of that policy by which it has been thus far aided, as well as the propriety, safety, and necessity of granting to it such additional assistance as may be required to insure its completion to the Pacific.

But while thus conceding the merits of other roads, I insist that, in regard to its length, its gradients, climate, its water communications and facilities for construction and operation, its agricultural and mineral resources, its relations to foreign and domestic commerce, its probable influence in promoting immigration, its effect upon the "Indian question," and upon the general development and prosperity of the country, the Northern road has merits and advantages which demand the prompt and efficient recognition of the Government.

ITS LENGTH.

This route seems to have been pointed out by nature as the line of communication between the waters of the two oceans. Toward the west the St. Lawrence and the great lakes reach half way across the continent, while on the same parallel of latitude, the Pacific, trending eastward, extends an arm to the great inland sea of Puget Sound, leaving a distance of only 1,775 miles between the waters of the two oceans. By the Central route, from San Francisco to Chicago, the distance from ocean to lake is not less than 2,400 miles, making a difference in favor of the Northern route of 650 miles. Governor Stevens has shown by carefully prepared tables that from New York city to the Pacific ocean, the difference in distance as between the Northern and Central routes is 420 miles in favor of the former, and that the average distance by this route, from Puget Sound to the principal ports of the Atlantic and the Gulf, is 316 miles less than the average distance from San Francisco to the same points; and that if equated distances are compared, the difference in favor of the Northern route will be increased 137 miles, making a total of 483 miles.

ITS GRADIENTS.

In its gradients, also, the Northern route has superior advantages. On this point I beg leave to read from the report of a select committee of this House, made in 1864: "The sum of ascents and descents from St. Paul to Seattle (on Puget Sound) is 21,787 feet against 29,387 on the Central. These figures give the best practical index of the effect of the gradients to increase the cost of transportation. Engineers allow one mile for every 52.08 feet of rise or fall, as denoting the additional working expense over a level route, which would add to the Northern route 412 miles, and to the Central route 556 miles."

The highest elevations on the Northern route are 5,330 feet, (Cadot's Pass,) and 3,000 feet at the Snoqualmie Pass. The greatest elevations on the Central route (already overcome by the Union and Central Pacific roads) are Evans' Pass, 8,242 feet; Rattlesnake Pass, 7,560 feet, and Bridger's Pass, 7,534 feet, making a difference between the two greatest elevations on either road, of 3,912 feet in favor of the Northern line. I make this comparison with the Central road, because that road having been constructed, demonstrates the practicability of building the Northern, where the gradients are far lighter.

And in order that my motives may not be misunderstood, I wish to state explicitly, that in the foregoing comparisons, and in all others that may be made of this with the Central route, I have selected the latter solely because of my desire to compare the Northern with one, the practicability and entire success of which is no longer doubtful.

CLIMATE.

The greatest obstacles which it was once imagined the Northern route would encounter, were the extreme cold and deep snows. That we have nothing to fear from these causes is now so well settled that discussion is unnecessary, and I will content myself with quoting a few items from the volumes of testimony on the subject.

Lieutenant Mullen, who spent four winters surveying and constructing a wagon road extending over a considerable portion of the Northern route, says:

"*The temperature of Walla Walla in* 46° *is similar to that of Washington city in* 38° *latitude; that of Clark's Fork in* 48° *to that of St. Joseph's, Missouri, in latitude* 41°, *that of the Bitter Root Valley in* 46° *is similar to that of Philadelphia in latitude* 40°, *with about the same amount of snow*, and with the exception of a few days of intense cold, about the same average temperature." * * * "From Fort Laramie to Clark's Fork I call this an atmospheric river of heat, varying in width from one to one hundred miles. On its either side, north and south, are walls of cold air, and which are so clearly perceptible that you always detect them when you are upon its shores.

"This affects the kingdoms of natural history, botany, and climatology to such an extent that herein we find mild winters, vigorous grasses, even in midwinter, that enable stock to be grazed upon the hills, and gives a facility for travel during the severest seasons of the year."

This remarkable meteorological condition is accounted for by the fact of the existence of an "infinite number of hot springs at the headwaters of the Missouri, Columbia, and Yellow Stone rivers," and hot geysers similar to those in California at the head of the Yellow Stone, together with the peculiar conformation of the Wind River mountains, and other mountain chains in that region, which "stand as a curvilinear wall to deflect and direct the currents of atmosphere which sweep across the continent. (By the bye, whence arises the name of the *Wind River* chain?) All their slopes are well located to reflect back the direct rays of the heat of the sun to the valleys that lie at their base. These valleys, already warm by virtue of the hot springs existing among them, receive this accumulative heat, which, driven by the new currents of cold air from the Plains, rises and moves onward, in the form of a river, towards the valleys of the Rocky Mountains, where it joins the milder currents from the Pacific, and diffuses over the whole country a mild, healthy, invigorating, and useful climate."

Mr. Philip Ritz, a gentleman who resides in Washington Territory, and who is represented to be entirely reliable, informs us that he has passed over this part of the mountains at various times in midwinter, and "so free from snow are these passes, from the effects of the wind currents coming from the great plains of the Columbia and the Pacific ocean, that I (he) found no difficulty in crossing at any time." He adds: "*The most surprising feature of the country is the small amount of snow that falls, and the fact that cattle winter here and keep fat on the natural pastures.* Mr. T. M. McCoy wintered two hundred head, within a few miles of the summit, without any food whatever except what they had by grazing on the dry grass." He states further that "there never has been a winter, with all their horses and men, and stations ten and twelve miles apart, (on the Central route,) that the Overland Company have carried the mail through on time; while on this route, from Hell Gate to Walla Walla, a distance of four hundred miles, and that a new route, with stations in places fifty miles apart, Messrs. Clark and Witcher carried the mail on horseback; and not only carried it through every trip (semi-weekly) during last winter, but were never once behind schedule time."

Governor Stevens says: "In the winter of 1854–5 the Flathead Indians passed through these passes in January, February, and March; whole tribes, with their women and children, and their pack animals laden down with meat and furs." Victor, head chief of the Flathead Nation, states that "since the memory of the Indian they had passed these mountains year after year through the winter months, the sole trouble being that there were some places where the snow was deep enough to cover up the grass, but in those places it was in the wooded portions, and two and a half feet was the greatest depth."

From Fort Benton eastward, the climate is somewhat more severe than on the western portions of the route. The coldest weather, however, and the deepest snows to be found on any part of the route, are in my own State, where it is well known they do not interpose any serious obstacles to the working of railroads. "The mean winter temperature of Fort Benton in 1853–4 was 25° 38′ above zero. At Montreal, on the Grand Trunk railroad, for the same year, it was 13° 22′, and for a mean of ten years 17° 80′ above zero. On the great Russian railroad from St. Petersburg to Moscow the comparison is similar. For a series of twenty-one years at Moscow the mean winter temperature was 15° 20′, and at St. Petersburg, for the same time, the mean was 18° 10′ above zero. At Fort Snelling, in Minnesota, the mean winter temperature of 1853–4 was 11° 46′, and the mean of thirty-five winters was 16° 10′ above zero. Thus, in the remarkably cold winter of 1853–4, Fort Benton was 15° warmer than Montreal, 11° warmer than Fort Snelling, 10° warmer than Moscow, and 7° warmer than St. Petersburg." The fact that railroads are operated without serious obstruction from cold and snow at Montreal, Moscow, and Fort Snelling is sufficient evidence that no difficulty will be encountered in the much warmer climate of Fort Benton.

"At Fort Benton and Fort Campbell, on the Upper Missouri, ever since they were established, some twenty-five years ago, the fur companies have taken their goods to their winter trading-posts, on the Milk and Marias rivers, in wagons, there not being snow enough for sleds."—*Letter of Governor Stevens, April* 30, 1860.

Prof. Hind, in an official report to the Canadian Legislature, says:

"That at the Red River settlement the mean winter temperature in 1856–7 was 6° 85′ above zero, and the total fall of snow for that year was 39.5 inches; when, for the same year, the fall of snow in Toronto, Canada, was 72.9 inches."

"The prevailing character of the winter months at this point are long-continued intense coal, with a clear, dry atmosphere." * * * * "The half-breeds, and of course the Indians, camp out in the open plain during the whole winter, and the only protection they enjoy consists of a buffalo-skin tent, and an abundance of buffalo robes."

Rev. John West, late chaplain to the Hudson Bay Company, remarks of the winter at the Red River settlement:

"The winter is nearly the same, as to the time it sets in and breaks up, as that of Montreal, but the frost is rather more intense, *with less snow and a clearer air.*"—*West's Journal*, p. 114.

The Red River settlement thus described lies more than two hundred miles north of the line of this road, and a thousand miles east of the mountain passes just mentioned, and is entirely beyond the reach of the peculiar meteorological influences which produce such remarkable effects there.

By Blodgett's rain maps it is shown that from St. Paul, Minnesota, to the Rocky Mountains, the total annual fall of snow averages only about *twenty inches.* In Canada, the average is *fifty inches.* In New England, from *seventy inches,* in the interior, to *one hundred inches* on the seaboard.

It is, therefore, obvious that neither the cold nor snows of the Northern route interpose any obstacles to the construction or successful operation of the road. Indeed, the snows are much deeper, and the cold more intense, on the elevated portions of the Central route than on the Northern, owing to the fact that the country about the headwaters of the Missouri and the Clark's branch of the Columbia is so much lower, and the air so much dryer, than in the passes of the Sierra Nevada and Rocky Mountains selected for the Central route, and to the additional facts that the mountain portions of the Northern route lie so near the Pacific as to be affected by the warm winds from that ocean, and by the "river of heat," described by Lieut. Mullen. The chief engineer of the Union Pacific road confirms this statement of the comparative mildness of the climate on the Northern route, by the declaration on page 10 of a report made in December, 1867, that the Oregon branch, which he advises his company to construct, "*avoids the Wasatch and Sierra Nevada ranges, with their troublesome snows and inhospitable winters.*"

On Fremont's Peak, in latitude 43⅓°, the limit of perpetual congelation was reached at an elevation of 10,000 feet. A large part of the elevated portions of the Central route is so near this limit as to account for the "troublesome snows" and "inhospitable winters," which it is sought to avoid by a line running further north, but through the warmer depression in the mountains. In fact, the *altitude* of the Central far more than counterbalances the *latitude* of the Northern line.

ITS WATER COMMUNICATIONS AND FACILITIES FOR CONSTRUCTION.

Considered with reference to its intermediate water communications, as well as to those at its termini, the northern route commends itself most forcibly to public favor. Passing westward from the head of Lake Superior, it crosses the Mississippi river at or near St. Cloud, Minnesota. Next, it strikes the Red river of the North, connecting with the navigation upon it, and thereby, with Lake Winnipeg and the Saskatchawan and Assinaboine rivers, and their tributaries, making over three thousand miles of navigable waters, and opening to settlement and to our commerce the great northern basin containing over 400,000 square miles of valuable territory. Thence it continues to the Missouri river, which it reaches at a point 485 miles distant from the head of Lake Superior. Thence it follows the general direction of the Missouri for a distance of 500 miles to the Great Falls near Fort Benton, to which point the Missouri is navigable. Thence to the navigable waters on Flathead and Clark's rivers—a distance of 200 miles. Thence along the navigable waters of the latter river, 175 miles. Thence to Lewis river, at the mouth of the Paluse, 140 miles, and thence down the valleys of the Lewis and Columbia to Portland. From this it will be observed that only 825 miles of road are necessary to be built at different points, in order to make a complete steam route from ocean to ocean. The advantages for construction afforded by these navigable streams cannot be over-estimated. No other route presents such engineering facilities, for it can be worked simultaneously from four different divisions, the extremity of each division resting on water lines, and thus the road can move on simultaneously on eight different sections—the longest division being the one from Fort Benton to the Columbia, and the longest distance of road, from a single point, accessible by water, being 300 miles. On every other route the distance between water lines will be 1,500 to 2,000 miles. This route follows the timbered valley of the streams nearly the whole distance, thus affording an ample supply of pure water, tember, and fuel. Starting, as it does, from the head of Lake Superior, the iron and other heavy materials for its construction can be brought to its own depot by water, as well as supplied at the various points of intersection of the navigable rivers on its line.

ITS AGRICULTURAL AND MINERAL RESOURCES.

The agricultural resources and mineral wealth of the country along the line of this road are not surpassed, if they are equalled, by any other part of the public domain. John Wilson, Third Auditor of the Treasury, and late Commissioner of the General Land Office, an eminent authority on all questions appertaining to the topography and resources of the country, says:

"*The lands along the Northern are all fit for cultivation*, and consequently the local business will support the road. *In fact, it is one of the most beautiful sections of country in the world*, and you travel comparatively a short distance west from St. Paul, till you feel the influence of the Pacific winds and strike the isothermal line of Southern Illinois, and where cattle can pasture the whole year. *The mineral wealth along the Northern is almost incalculable.*"

Governor Isaac I. Stevens, who surveyed the entire route, and spent several years in exploring the country contiguous to it, says:

"In the whole distance from Breckenridge to Seattle, a distance of 1,544 miles, the route passes through only about 320 miles of uncultivatable country. East of Breckenridge to St. Paul and Lake Superior, the country is exceedingly rich and inviting to the settler."

Lewis and Clarke, under date of May 17, 1805, and while encamped on the Rocky Mountains, state in their journal: "The country along the Rocky Mountains, for several hundred miles in length and about fifty wide, is a high, level plain, *in all its parts extremely fertile.* This plain is chiefly interrupted near the streams of water, where the hills are steep and lofty, but the soil is good. Nearly the whole of this wide-spread tract is covered with a profusion of grass and plants, *which are, at this time, as high as the knees.*"

Grass and plants as high as the knees on the 17th of May! Gentlemen from New England and the northern parts of the Middle States will please note this *date.* But the agricultural character of the northern part of our own territory is so well known that it is unnecessary to adduce further evidence. There is, however, upon our northern border, a vast territory capable of making eight States the size of New York the whole of which would be tributary to this road about which there is less general information, and to which I may briefly refer.

The English explorer, Hind, writes of this country:

"Of the valley of Red river, I find it impossible to speak in any other terms than those which may express astonishment and admiration. I entirely concur in the brief but expressive description given to me by an English settler on the Assinaboine, that *the valley of the Red river, including a large portion belonging to its great affluents, is a paradise of fertility.*" * * * "The Saskatchawan, which gathers its waters from a country greater in extent than the vast region drained by the St. Lawrence and all its tributaries from Lake Superior to the gulf, is navigable by either the north or south branch for more than a thousand miles of its course. *So mild is the climate on the south branch of this river*" (200 miles north of the line of this road) "*that the Indians hunt the buffalo on horseback all winter, and so little snow is said to fall, that snow-shoes are seldom used.* That this extensive territory drained by the Saskatchawan and its tributaries is fit for settlement, in as far as regards climate, is fully proved by the success which attends the farming operations which are carried on, though on a small scale, at the various trading posts, and by the fact that cattle and horses at these establishments are generally left to forage for themselves during the winter. As regards the soil, from what is yet known of the country, there is not, perhaps, on the globe, so great an extent of country so little broken by barren tracts. *Regarding the territory in its general aspect, there is not, in the universe, a finer field for colonization. It has a salubrious climate, and the soil, in many places, as at Red river, is unsurpassed for fertility. Iron ore, lead, and salt are to be found in abundance,* and the whole territory, from Lake Winnipeg to the base of the Rocky Mountains, is intersected by navigable rivers and lakes."

He mentions some specimens of farming on the Assinaboine river, and says that one farmer had "grown 56 measured bushels of wheat to the acre. Forty bushels to the acre is a common return on new land. His turnips were magnificent; four of them weighed 72 pounds. A portion of the potato crop was still in the ground. They far surpassed in quantity, quality, and size, any I had ever seen before. I counted 13, 14, and 16 potatoes, averaging 3½ inches in diameter, at each root. * * * "Indian corn, if properly cultivated, and an early variety selected, may always be relied on. The melon grows with the utmost luxuriance without any artificial aid, and ripens perfectly before the end of August."

This fertile district, which is destined, at no distant day, to be the wheat granery of the world, contains an area of about 400,000 square miles, the whole of which will be tributary to the Northern road, as will readily appear upon an examination of the map of that region. The revenues of the road will not only be immensely increased by the commerce of that great country, but its construction within our own territory will prevent any attempt to build one by the British Government, and will guarantee the speedy annexation of British Columbia to the United States without conquest or purchase.

Returning to an investigation of the mineral wealth of the States and Territories on the Northern road, I find it stated by the mining commissioners of the Treasury Department, J. Ross Browne and J. W. Taylor, that the product of gold and silver for the year 1867 was as follows:

Montana	$12,000,000
Idaho	6,500,000
Washington	1,000,000
Oregon	2,000,000
	$21,500,000

It will be remembered that the delegates from Montana and Idaho protested from their places on this floor against the order to print J. Ross Browne's report, on the ground of the injustice alleged to have been done their Territories, by the great undervaluation of the product of gold and silver. The entire product of the precious metals in Oregon and the said Territories since 1862 is estimated as follows:

Montana	$65,000,000
Idaho	45,000,000
Washington	10,000,000
Oregon	20,000,000
	$140,000,000

All this has been done at enormous cost and great disadvantage. Mining in Montana and Idaho has hardly begun. They are, doubtless, immensely rich in minerals, but so long as it costs $600 per ton to transport mining machinery there, we may not expect a thorough development of this wealth.

But notwithstanding all these advantages of distance, gradients, climate, soil, water communications, and mineral wealth, the question still remains, "Is it *safe*, *expedient*, and *necessary* for Congress to grant the proposed loan of national credit at this time?"

IS IT SAFE?

It is, and for these reasons: The bill provides that the issue of Government bonds shall *ipso facto* constitute a mortgage upon the road and all its appurtenances; and upon the failure or refusal of the company to pay the interest, or to redeem the bonds or any portion of them when due, the Secretary of the Treasury is to be authorized to take possession of the road, with all the rights, functions, immunities, and appurtenances thereto belonging, and also all the lands granted to said company which at the time of such default shall remain in their ownership, and hold the same for the full indemnity of the United States. The Secretary of the Treasury is also required by the Pacific Railroad law to retain one half of the compensation for services rendered by the road, as a fund for the redemption of the principal of the bonds; and also, as an additional sinking fund, five per cent. on the net earnings of the road. It can be shown conclusively that if the Government should actually donate the entire amount of bonds asked for, and thereby secure the construction of the road, it would be a most profitable investment for the United States; but fortunately the Pacific roads already partially built demonstrate the fact, that the Government will never be called upon to pay one dollar for them.

The President of the Kansas Pacific road states officially that "the Government is receiving directly into its treasury from that company more than the interest on all the bonds issued to it, added to a sinking fund that will extinguish the principal at maturity. This is altogether irrespective of the economy arising from the substitution of rail for wagon road in the transportation of mails, troops, and supplies for the Government."

The Committee on Military Affairs of this House, to whom was referred the letter of Lieutenant General Sherman, dated March 4, 1868, recommending Government aid to extend the Kansas Pacific road, as a "military necessity" and a measure of public economy, reported the following facts and conclusions drawn from official records:

The cost to the Government for transportation on the Union Pacific Railway, Eastern Division, in 1867, was	$511,908 24
If the military supplies had been wagoned, and the mails carried by stage, and the troops marched, (taking the average rates at which the Government made its transportation contracts for that year, as shown by certificates of the Departments of the Quartermaster General and Postmaster General,) the cost would have been	1,358,291 02
Saving to the Government in 1867	$846,382 82

"*At this rate of saving,*" (say the committee,) "*all the United States bonds issued in aid of this road, principal and interest, would be extinguished in less than four years.*"

It appears, therefore, by this report, and from official records, that from the first year's actual business of this unfinished road (its average length being only 201 miles) the company have paid all the interest due on the Uuited States bonds issued to aid it; that they have deposited the necessary sinking fund, and that, in addition, it has saved to the Government, in a single year, over twenty-five per cent. on the entire amount of aid it has received. It should be borne in mind, also, that this saving will be much greater when the road shall be completed, and that it will not be for a single year, but for all time.

Equally favorable results have been produced by the Union Pacific road, as will appear from the following statement, compiled from the official records of the Quartermaster General's office. In 1866, wagon transportation over the route now occupied by this road cost an average of 28.4 cents per ton per mile; in 1867, an average of 39.4 cents per ton per mile; while for the season from January 1, 1868, to March 31, of the same year, the lowest contract that could be obtained, was 50 cents per ton per mile. The cost of Government transportation over this road is 10.5 cents per ton per mile. The amount paid by the Quartermaster General to this railroad in 1867, for Government transportation, was $699,698.81. Had this same freight been transported by wagons at the contract price for that year, (39.4 cents,) the cost would have been $2,625,536.41. So that *the money actually saved to the Government in one year, with the road in operation an average distance of only* 386 *miles, was* $1,925,837.60. This is the actual *profit* to the Government, for the company has paid into the United States treasury the full amount of interest due on the bonds received, and also the per centage of sinking fund required by law.

The "Central Pacific," I am informed, presents an equally good balance sheet as between the company and the Government, but I have not the official figures at hand. It appears, therefore, that from two of these roads, having an aggregate of only 587 miles completed, *the Government received a net profit of* $2,772.140.42 *in one year*. The saving made by the Central Pacific will doubtless swell this aggregate *net profit to the sum of* $3,500,000 *for* 1867; *making an actual net profit large enough to pay, in five years, the entire amount of Government bonds loaned for the construction of these portions of said roads*. The business necessity for all these roads is clearly demonstrated by their extraordinary success while yet in an unfinished condition. I know nothing in the history of railway construction that is at all comparable to it. Nor can we avoid the conclusion, that these favorable results to the Government will be largely augmented. when the roads shall be completed. Their wonderful success demonstrates, also, the statement already made, that the Government will never be called upon to pay one dollar for them. The companies owning them will at no distant day become among the richest private corporations in the world, and it is by no means probable that they will, by default in payment, permit the Government to foreclose her mortgages, and take possession of this immensely valuable and productive property for one-third of its cost. If they should do so, even that would be a transaction by which the nation would gain many millions of dollars. The same reasons which insure the safety of the Government in regard to the roads just referred to, apply with equal, if not greater force, (as we shall hereafter see) to the Northern Pacific. The amount of the loan asked for is about $50,000,000, which is to be a second mortgage, the first being for the same amount.

Mr. Edwin F. Johnson, chief engineer of the road, and one of the safest, most reliable, and intelligent engineers in the country, officially estimates the entire cost of construction and equipment at $156,847,500. That he has probably not overestimated this cost will appear from the letter of the Secretary of War in answer to the Senate's resolution of July 24, 1866, in which he transmitted to the Senate reports from A. A. Humphreys, Chief of Engineers, and General J. H. Simpson, Lt. Col. of Engineers. The former officially estimates the entire cost of construction and equipment at $215,306,208; the latter estimates it at $206,600,340.

Hence, though the Government shall take a second lien, it is rendered secure by the fact that the company will have to expend, of their own money, from fifty to a hundred millions of dollars, in addition to the amount of the liens of the first mortgage bondholders, and of the Government, which amount thus to be expended by the company will be subsequent to the lien of the United States. It is therefore a transaction which, on the part of the Government, is *absolutely safe.*

IS IT EXPEDIENT UNDER EXISTING CIRCUMSTANCES?

When I entered upon the investigation of this subject, the questions that arose in my mind were; "Are we as a nation sufficiently careful and discreet?" "Are we not in danger of going too fast?" "May we not be over-confident of our powers, and too ready to undertake great and costly enterprises, from which older and more experienced peoples, with so heavy a burden of debt, would shrink?" With these conservative queries in mind, I began the comparison of the United States with Great Britain and France, in order that I might profit by the wisdom of the statesmen who control the destinies of those great nations. I earnestly solicit the attention of the House to a statement of the results of that comparison.

Since 1860, Great Britain has aided the construction of 4,600 miles of railroad in India, at an estimated expense of $440,000,000, for which sum the credit of the Imperial Government was granted to private companies, guaranteeing five per cent. interest on all money invested in Indian railroads. The public debt of Great Britain in 1863, at the time she was thus loaning her credit for the construction of said roads, was $3,915,000,000, and her population a little less than 30,000,000, making a debt to each person of $130.40. I speak, of course, of Great Britain proper, for it is well known that her dependencies are not sources of any considerable net revenue.

The standing army of Great Britain, in 1863, on a "peace establishment," was 198,518 officers and men, and 27,331 horses. The current expenses of the British Government that year were about the average of other years of peace, and amounted, when reduced to the basis of our currency, to $469,000,000.

Within the last ten years France, also, has loaned her credit for the construction of railroads to the amount of $620,000,000, which was authorized to be raised by private companies by debentures, "on which the Government guaranteed 4 per cent. interest, and .65 sinking fund, for the paying them off in fifty years." The public debt of France in 1862, was $2,206,000,000, and her population 37,000,000, making a debt to each person of $59.65. The standing army of France, as estimated for in 1863 was, "peace establishment," 404,195 men, and 86,368 horses; "war establishment," 757,725 men, and 143,238 horses. The current expenses of the French Government for that year, on a peace basis, were, when reduced to the standard of our currency, $515,900,000.

The public debt of the United States, as officially reported by the Secretary of the Treasury, was, on the 1st of September, 1868, $2,535,614,313, and our population is about 40,000,000, giving, as the indebtedness of each person, $63. The standing army of the United States does not exceed, I believe, 40,000 men, and our current expenses, including payment of interest on debt, are within $350,000,000 per annum. Our army, therefore, is only about one-fifth as large as that of Great Britain, and less than one-tenth that of France. *The expenses of Great Britain exceed ours by* $119,000,000 *per annum; those of France are greater than ours by* 165,000,000 *per annum.* Consider, now, the probable resources of these three great nations, on which the future may rely to meet their financial burdens. Their past increase in population and in wealth will furnish the surest guide to a safe estimate for the future. On this point I submit a table showing the actual increase in population of the United States by decades, from

1790 to 1860; of Great Britain, from 1793 to 1861; of France, from 1801 to 1861:

YEAR.	UNITED STATES.	GREAT BRITAIN.	FRANCE.
1790	3,929,827	....	....
1793	....	14,500,000	....
1800	5,305,937	16,000,000	....
1801	....	....	27,349,000
1810	7,239,814	....	....
1812	....	18,000,000	....
1820	9,638,191	....	....
1823	....	21,193,438	....
1821	....	....	30,461,000
1830	12,866,020	....	....
1833	....	24,304,799	....
1831	....	....	32,569,000
1840	17,069,453	....	....
1841	....	27,041,031	34,230,000
1850	23,191,876	27,300,000	....
1851	....	....	35,283,000
1860	31,445,000	....	....
1861	....	29,334,788	37,400,000

Increase in the United States for 70 years, 700.41 per cent.
Increase in Great Britain for 68 years ......102.30 "
Increase in France for 60 years .... 37 "

It will be observed that the percentage of gain by the United States over Great Britain and France, has been very much greater during the later decades than during those at the beginning of the century. For instance, from 1800 to 1810, this increase in the United States was 36 per cent.; while from 1800 to 1812, the increase in Great Britain was 12 per cent.; the percentage of gain in the former, being *three times* as great as in the latter nation.

The increase in the United States from 1850 to 1860 was 35 per cent., and in Great Britain for the same period it was only $\frac{70}{100}$ per cent., making a per centage of gain in the United States for that decade, *fifty times* greater than in Great Britain.

This comparison will disclose still more remarkable results as between the United States and France, but my time will not permit its further extension.

I beg the attention of the House to another table showing the increase in wealth in the United States by decades, from 1790 to 1860, and the estimated increase in Great Britain at various periods from 1793 to 1860, compiled from the "Annual Record for 1864:"

YEAR.	Value of all kinds of property in the United States—estimated.	Value of all kinds of property in Great Britain.
1790	$750,000,000	....
1793	....	$7,132,000,000[a]
1800	1,072,000,000	8,753,400,000[b]
1810	1,500,000,000	....
1812	....	10,212,300,000[c]
1816	....	10,400,000,000
1820	1,882,000,000	....
1823	....	10,698,600,000[d]
1830	2,653,000,000	....
1833	....	17,199,458,400[e]
1840	3,764,000,000	....
1841	....	19,452,000,000[f]
1850	7,135,780,000	22,564,320,000
1858	....	29,178,000,000
1860	16,159,000,000	....

(*a*) Joseph Lowe. (*b*) Sir W. Pulteney. (*c*) Colquhoun. (*d*) Joseph Lowe. (*e*) Pebrer. (*f*) G. R. Porter.

In my printed remarks, I will annex to each estimate of the wealth of Great Britain, the name of the statistician upon whose authority it is stated. You will observe that during the period of twenty-five years from 1833 to 1858, the increase in Great Britain was 66 per cent.; and in the twenty years from 1840 to 1860, the increase in the United States was over 330 per cent. From 1800 to 1858, (58 years,) Great Britain's increase was 233 per cent., while from 1800 to 1860 (60 years) this nation's increase was 1,400 per cent. During the 40 years from 1793 to 1833, the increase of wealth in Great Britain was *one hundred and fifty-one per cent.* During substantially the same period, viz: from 1790 to 1830, the United States' increase of wealth was *two hundred and fifty-three per cent.* From 1833 to 1858, Great Britain's increase in wealth was *sixty-nine per cent.* From 1830 to 1860, (the railroad era,) the increase of wealth in the United States was *five hundred and eight per cent.*

It will be seen that, during the first period, including the infancy of our country, when our *percentage* of increase would be supposed to be greatest, our gain was *less than double* that of Great Britain; while, during the thirty years from 1830 to 1860, (since we entered upon railroad development,) it has been more than *seven times* that of Great Britain.

Compare the growth of the two countries in their annual products:

In 1858, the products of Great Britain were	$2,917,800,000
In 1833, " " "	2,503,564,248
Increase in 25 years	414,235,752
Percentage of increase in 25 years, 17.	
In 1860, the products of the United States were	$3,804,000,000
In 1840, " " " "	1,063,135,000
Increase of annual products in 20 years	2,740,865,000
Percentage of increase during same time, 256.	

Gain in the value of annual products during this period, more than *eighteen* times greater than in Great Britain.

Take one other period.

In 1800, the value of Great Britain's products was $1,313,000,000, which, deducted from that of 1858, shows an increase in 58 years of $1,604,800,000, or 122 *per cent.*

In 1800, the value of the products of the United States was $300,300,000, and in 1860, $3,804,000,000; making an increase in 60 years of $3,504,700,000. Gain during this period, *nine* times greater than in Great Britain.

From a comparison of the gains of these various periods, it will be observed that, during the later years, it has been very much greater relatively in the United States, than during the earlier part of the century, *thus proving that this vast disparity in our favor is likely to increase, rather than diminish, in the future.*

The relative ability of the two nations to bear their respective burdens may be seen in the light of another fact. When our war ended, three and a half years ago, the entire debt, liquidated and unliquidated, amounted to over $3,300,-000,000; and on the 1st of last September it was only, in round numbers, $2,500,000,000—*a reduction by actual payment in three and a half years of* $800,000,000. In 1816, the debt of Great Britain was $4,200,000,000; and in 1863 it was, in round numbers, $3,900,000,000—a reduction of only $300,000,000 in *forty-seven years.*

I have not the data, nor the time to extend these comparisons as to increase of wealth to France, but they would not, I think, be very dissimilar from those just made between the United States and Great Britain.

I have presented these somewhat voluminous statistics for the purpose of showing, that we are far more able than either Great Britain or France, to pay the debt resting upon us, and to incur such further liabilities as may seem necessary for the development of our country; and also for the purpose of introducing the example of those old conservative nations to refute the arguments of those who say, "now is not the time; we are too much in debt to undertake any of these great enterprises."

Great Britain, with a debt much larger than ours; with a smaller population; with a far less ratio of increase in population, wealth, and annual products; with a standing army at least five times as great as ours to support; loans her credit to the amount of $440,000,000 to build 4,600 miles of railroad, for the purpose of developing one of her remote dependencies; and yet we pause and question our ability, when it is proposed to loan the credit of our Government to the extent of only about fifty millions, in order to insure the construction of *two thousand miles* more, of railroad, which will develop, in the very heart of the Great Republic, the incalculable wealth that now lies but half concealed in its vast prairies and mountains. If it be prudent for Great Britain to do so much for far distant India; if France, with a debt almost as large as ours; with a population that in sixty years has increased only one-twentieth as rapidly as ours, with a standing army of from four hundred thousand, to seven hundred and fifty thousand men to support—if France, under these circumstances, acts wisely in granting her credit to the extent of $620,000,000 in order to develop the resources of a country not larger than one of our States; shall it be said that we are injudicious, and reckless of the nation's credit, when we propose a loan of only about $50,000,000 to open up a great highway for the world's commerce, and to develop the wealth of five hundred thousand square miles of the richest mineral and agricultural territory on the globe?

Why, sir, since I began the investigation of this question, I have been astonished to find, that with incalculably greater inducements to engage in these giant enterprises of the century, we are far behind the other nations in the encouragement given to them. Young America is really old fogyish on this point. The wise statesmen of the Old World see that in the full development of their national resources, consists the ability to meet their national obligations. Shall we profit by their example?

IS IT NECESSARY AT THIS TIME?

It is, for several reasons.

1.—*The road is a "military necessity."* All our efforts thus far, to secure permanent peace with the Indian tribes, have proved disastrous and costly failures. The treaty system has been tried until its palpable absurdities are apparent. Treaties have been so often violated by both parties, that the Indian, at least, has come to regard them as binding only so long as necessity or self-interest compels obedience. Without stopping now to ask which party is most at fault, I will content myself with the assertion of the fact so often demonstated within the last twenty years, that lasting peace cannot be obtained by the treaty system.

Can it be secured by war? Not with the facilities we now have for conducting such wars. It is often said that the Indians should be turned over to the War Department, and annihilated if they will not cease their outrages. It is not difficult to turn them over, by law, to the military authorities, but for the military authorities to get possession of their property is quite another matter. Catching the hare is in order before eating him. And as to the proposition to "annihilate them" often so glibly uttered, it will, when tried, be far more likely to annihilate the treasury than the Indians. I hold in my hand a report from the Secretary of War, dated December 30, 1866, which states that the attempt to suppress Indian hostilities in 1864 and 1865 cost $30,530,942.93. An estimate, made at the time, of the number of Indians killed by those expeditions, showed about one dead Indian for every $2,000,000 expended. At that rate of expenditure, the "annihilation" of two thousand Indians will cost as much as the suppression of the rebellion.

The Indian war expenses of 1864 and 1865 certainly did not exceed, if they equalled, the average expenses for that purpose during the last six years. For several months in 1866, the cost, (as stated by General Grant,) was about one million dollars per week. The total expenditure, therefore, for Indian wars, during the last six years, will doubtless largely exceed $100,000,000.

For all this vast expenditure, what has been done toward securing a permanent peace? Absolutely nothing. The savages are far more turbulent and troublesome than they were six years ago. Our dispatches from the Plains are still burdened with accounts of all sorts of plundering, mail-robbing, scalping, and murdering. For every Indian killed, a dozen soldiers have been sacrificed. The experiences of the last six years, therefore, are not such as to justify a very strong confidence in the "annihilation," or even in the thorough chastisement of the savage, by our present mode of warfare. I believe in punishing him when he does wrong, and I am entirely satisfied that in no other way can we have peace. He regards no compact that he does not fear to break. He knows no law but the law of force. He respects no authority but that in which he sees the power to enforce obedience.

It is no fault of the army, or of the Generals in command, that we have so utterly failed. The army is a portion of the same that won imperishable glory on the bloody fields of the South. The commanders are among those most renowned for tact, dash, and energy. The names of Sherman, Sheridan, and Hancock, are a sufficient guarantee that nothing within the range of possibilities has been left undone. Why, then, have we so signally failed? The answer is found in the inherent difficulty of the situation, viz: *the impossibility of "getting at" the foe.* Infantry are worse than useless. Cavalry scour the plains for a few hundred miles from head-quarters, but do not find the enemy. A million and a half square miles of unsettled prairie and mountain fastnesses offer him a sure and safe retreat. A score of these wild horsemen of the Plains, accustomed to subsist by the chase, and mounted on their fleet ponies, with no baggage but a gun and blanket, with no tents to pitch and no base of supplies to guard, can bid defiance to whole regiments of cavalry. The Indian system of warfare is to divide into small predatory bands, strike a blow at some defenceless frontier settlement, plunder and murder some emigrant train, or rob the United States mail, and before our troops are aware of their outrages, they are hundreds of miles away in the interior, or at some other remote quarter, committing new depredations. It is this immunity from danger, this certain and ever open door of escape from punishment, that makes the Indian so troublesome; and it is the impossibility of finding him, that makes all our military expeditions such utter failures.

So long as he has behind him this vast unoccupied territory, into which he can flee and be safe, the Government will be powerless to enforce his obedience, and will continue to squander untold millions in fruitless efforts to conquer a peace. What, then, is the remedy? There is but one—the construction of railroads across the continent, whereby troops can be speedily moved, and self-protecting settlements be made.

All of our generals who have commanded in the Indian country, and whose opinions are most valuable on this subject, agree that this is the true solution of the "Indian question."

In his report as Secretary of War *ad interim*, General GRANT says:

"During the last summer and summer before, I caused inspection to be made of the various routes of travel and supply through the territory between the Missouri river and the Pacific coast; the cost of maintaing troops in that section was so enormous that I desired, if possible, to reduce it. This I have been enabled to do, to some extent, from the information obtained from these inspectors; but, for the present, the military establishment between the lines designated must be maintained at a great cost per man. The completion of the railroad to the Pacific will materially reduce this cost, as well as the number of men to be kept there. *The completion of those roads will also go far towards a permanent settlement of our Indian affairs.*

On the 4th of March, 1868, Lieutenant General SHERMAN, writing in regard to the Kansas Pacific road, says:

"*This road is a military necessity,* and will very much stimulate the settlement of that region of our public domain."

Major General SHERIDAN wrote, on the 2d of May, 1868, concerning the same road, as follows:

"I know that pecuniarily it would be to the advantage of the Government to help this road." * * * "*But, in addition, it almost substantially ends our Indian troubles by the moral effect which it exercises over the Indians, and the facility which it gives to the military in controlling them.*" * * * "No one, unless he has personally visited this country, can well appreciate the great assistance which this railroad gives to *economy, security, and effectiveness in the administration of military affairs in this department.*"

Major General HANCOCK, under date of June 4, 1868, says:

"From my personal experience, obtained while I was on the Plains, with respect to the transportation of troops and supplies by the railroad in question, (the Union Pacific, E. D.,) as well as its great importance in connection with the settlement of the country, I feel that I can speak in strong language as to the necessity of this road being extended as rapidly as practicable to Fort Wallace, where I have lately been."

Major General SANBORN, a member of the peace commission, and a gentleman of large experience in Indian affairs, on the 31st of October last, wrote:

"*Railroads, more than all other things, extend our civiliza-*

tion over men and remote regions, and will do more in a single decade to civilize Indians, and to compel them to abandon nomadic habits, than could be done in a century without them. The members of the commission, so far as I know, are all advocates of two more lines of road to the Pacific."

Precisely the same reasons, on which these eminent generals base their recommendations of aid to the Kansas Pacific road, as a "military necessity," apply with equal or greater force to the Northern Pacific. Between the 44th degree of north latitude (a line about half way from the central to the northern routes) and our northern boundary, there resides over one hundred thousand Indians, being one-third of the entire Indian population of the West. Of this number, thirty-five thousand are Sioux, confessedly the most warlike and dangerous of all the tribes. Some fifty thousand more belong to the Oregon, Washington, Montana, and Idaho tribes, with many of whom we have been in a state of chronic warfare for years. In British America, adjacent to our northern boundary, are thirty-five thousand more, who, in the event of war, readily cross the line and join the tribes on this side.

Now, including British America, there is a vast unoccupied country north of the line of the Union and Central Pacific roads, at least a thousand miles in width by twelve hundred miles in length, into which, when pursued, the savage can retreat. Twelve hundred thousand miles of prairie, wilderness, and mountain solitude, for a place of refuge! Surely he can desire no more certain immunity from punishment.

So far as these warlike northern tribes are concerned, the construction of only one road across the continent will have the effect to irritate, without furnishing the means to control them.

They will sweep down upon the line of the Union Pacific, murder and scalp a train of passengers, and escape unharmed to their northern refuge, where it will still be almost as impossible to "get at" them as it is now. The cheapest, and, in fact, the only way, to protect the road now nearly finished, will be to construct two more. Complete the Central, Northern, and Southern roads, and three lines of settlements, not more than from four hundred to six hundred miles apart, will span the continent. Towns and cities, capable of self-protection, will spring up at convenient distances ; troops can be speedily transported and cheaply subsisted. War may then be made effective, if war shall be necessary. But the fact that it can be effective will be apparent to the Indian, and will soon teach him that in obedience is his only safety. He will then peaceably submit to the only law he recognizes—the law of force—and our Indian wars will be "substantially ended."

2.—*It is necessary as a measure of retrenchment.*

In the words of General Grant, "*the construction of a road by the proposed route would be of very great advantage to the Government pecuniarily.*"

But I can present no argument on this point half so forcible as the quotation of the entire letter of that far-seeing statesman and great soldier :

HEADQUARTERS ARMY OF THE UNITED STATES,
April 20, 1866.

The construction of a railroad by the proposed route would be of very great advantage to the Government pecuniarily, by saving the cost of transportation to supply troops, whose presence in the country through which it is proposed to pass is made necessary by the great amount of emigration to the gold-bearing regions of the Rocky Mountains. In my opinion, too, the United States would receive an additional pecuniary benefit in the construction of this road, by the settlement it would induce along the line of the road, and consequently, the less number of troops necessary to secure order and safety. How far these benefits should be compensated by the General Government beyond the grant of land already awarded by Congress, I would not pretend to say. *I would merely give it as my opinion, that the enterprise of constructing the Northern Pacific railroad is one well worth fostering by the General Government, and that such aid could well be afforded as would insure the early prosecution of the work.*

U. S. GRANT,
Lieutenant General.

Aside from the testimony of this eminent authority, the facts will demonstrate that the prompt construction of the Northern Pacific road is a necessary measure of retrenchment. We have just seen that the cost of our fruitless Indian wars has reached the enormous sum of $100,000,000 in six years. The papers inform us that Mr. A. H. Jackson, of New York city, has compiled from the official records, and will soon publish, an exhibit of the cost of the several Indian wars during the last thirty-seven years, which will show the loss of over 20,000 lives, and an expenditure of more than $750,000,000, making an average annual loss during that time of more than 600 soldiers and $20,000,000. I have not been able to make a minute personal examination of these records, but having been chairman of the Committee of Indian Affairs for six years, I have given much attention to this subject, and I am satisfied Mr. Jackson's statement is not an exaggeration. But in order that I may be indisputably within the bounds of truth, I will place the estimate of expense in this regard at the average, of $15,000,000 per annum for the last thirty-seven years.

The highest military authorities of the nation have unanimously advised us that in our Pacific railroad system will be found the peaceful solution of the "Indian question." In the words of General Grant, in his official report as Secretary of War *ad interim*, "*the completion of these roads will go far toward a permanent settlement of the Indian question*;" or, in the still more emphatic language of General Sheridan, just quoted, they will "*almost substantially end our Indian troubles.*"

Says General INGALLS:

"*Nothing would so much conduce to the settlement of the difficulties with the Indian tribes as the facilities afforded by railroads for a speedy communication through this country.* In my opinion, from an experience of many years in the quartermaster's department in the West and Northwest, *it is of the utmost importance to the nation that this road* (the Northern Pacific) *should be constructed at the earliest moment possible.*"

Extract from letter of Gen. INGALLS, dated April 25, 1866.

In April, 1866, Quartermaster General MEIGS wrote of the Northern Pacific road:

"*The enterprise is one worthy of the nation. As a military measure, contributing to national security and defence alone, it is worthy the cost of effectual assistance by the Government.*"

The entire Indian population of the West numbers about 300,000, one-third of whom, as already stated, reside north of latitude 44°. As to the comparative expense of wars against them and the Southern and Central tribes, I request the clerk to read from an official report of the Quartermaster General, in response to a resolution offered by myself on the 7th of June, 1866. I have italicised the items which are in whole or in part chargeable to the Northern tribes.

EXPEDITION.	Amount for each.	Total per year.
1864.		
Northwestern Indian Expedition	$5,236,000	
Apaches	737,908	
Walnut Creek	73,487	
Smoky Hill	83,341	
Little Blue	23,786	
North Platte	213,810	
California and *Oregon*	1,018,034	
Plum Creek	80,495	
Sand Creek	182,524	
Kiowas	311,617	
Republican River	47,596	
New Mexico	101,774	
		$9,110,372
1865.		
Northwestern Expedition	$1,394,190	
New Mexico	298,349	
Districts of the Plains	13,470,957	
Kiowa Expedition	100,703	
California and *Oregon*	1,553,816	
Republican River	30,173	
Northwest Territory	2,415,168	
		19,263,856
Total for the two years	....	28,374,228

Of the above items, those designated, respectively, "Northwestern Indian Expedition," "Northwestern Expedition," and "Northwest Territory," are all, manifestly, for expeditions against the northern tribes. To those items may be added one-half of the sums mentioned for California and Oregon, though much the largest part was doubtless expended in the latter State. I add, also, only one-third of the item for the "District of the Plains," though the detailed statements of said report will show that this will fall below the actual proportion chargeable to these tribes. It appears, therefore, that of $28,374,228 expended in two years, on account of the quartermaster's department, $14,888,268, *or more than* 50 *per cent. of the whole* was for military expeditions against the Indians living north of latitude 44°. I have no official statement of the comparative costs for other years since 1862, but the proportion to be charged to the Northern Indians will exceed that for the years mentioned in said report.

It may, therefore, be safely asserted that more than one-half of the $100,000,000 expended during the last six years, is properly chargeable to the futile efforts against Indians on, or near, the line of the Northern Pacific road. I will, however, estimate the proportion thus chargeable to them during that time at 40 per cent. of the whole. Had the three roads been completed six years ago, they would have "substantially ended our Indian troubles," and the proportion of saving that could have been credited to the Northern Pacific, would have been at least $40,000,000 during that time, or at the rate of over $6,500,000 per annum.

That my estimate of the retrenchment which would have been effected by this road is below the actual amount will conclusively appear by reference to the following facts, substantiated by official records, viz: The number of military posts now established on the line, and in the vicinity of this road, is 28. The entire number of troops about 4,500. The amount of supplies necessary for their subsistence is 22,995 tons per annum. The cost of transportation alone of these supplies for the year 1867, under the contracts of that year, estimating the distances as officially stated by the Quartermaster General, was $6,158,972, to which must be added the pay of the soldiers and officers, and other large incidental expenses, which will swell this sum to nearly $8,000,000, making almost one-sixth of the entire amount of the Government loan asked for to construct the road.

It is stated, on good authority I believe, that a cavalry force of 5,000 men, under command of General Sheridan, is now moving in pursuit of the Indians. This, including the troops at the various posts, is a larger force than was employed in the summer of 1867, when General Grant estimated the expenses at $1,000,000 per week. Winter is also a far more expensive season than summer for army movements on the prairies, and in the mountains. May we not, therefore, safely estimate the present current expenses of Indian wars at $1,000,000 per week—$144,000 per day? And it must not be forgotten that the money thus expended is a total loss, while in the case of the contemplated railroad aid it is a mere loan of credit, requiring no actual outlay. Let it be remembered, also, that this vast fruitless drain upon the treasury, ranging from $15,000 to $144,000 per day, as just shown, is to continue indefinitely until the true remedy is applied. Shall we apply it, and thereby lighten the burdens of the people, or, under the false pretence of economy, continue the profligate, "penny-wise and pound-foolish policy of the past?"

A very great additional saving, amounting to many hundreds of thousands of dollars per annum, would be effected by the construction of this road, in mail transportation to and from the Northwestern States and Territories.

But one of the most important economical features of this measure may be seen in the facilities it will afford to the people of the Northwest, for retrenchment in their business expenses, and in the diversion of labor, it will effect, from present unproductive channels into pursuits that will add to the wealth of the nation.

During the year 1865, the aggregate shipments of twenty-seven firms, from the town of Atchison alone, amounted to

21,531,830 pounds of assorted merchandise, requiring for its transportation 4,917 wagons, 7,154 mules, 27,675 oxen, and employing 5,256 men. The total amount of capital invested in wagons, mules, oxen, horses, &c., employed in that trade was over $6,000,000. It is the opinion of those best informed upon the subject that this trade has doubled every year since 1865. Add to this the shipments over the plains from all the other points, and what an army of men, horses, wagons, &c., and what a vast amount of capital, may, by the construction of railroads, be discharged from this unproductive business, and employed in some pursuit that will increase the wealth of the country.

IT IS A NECESSARY REVENUE MEASURE.

In other words, the national credit and national revenues will be greatly benefitted by the construction of the road, and for reasons which will appear from the following considerations:

1. *It will stimulate immigration.*

One of the chief causes of the wonderful increase of the United States in population and wealth, as already shown, is our unparalleled immigration. There is nothing in the history of nations to which it can be compared. The causes which impel this vast tide of humanity to our shores are of course various, but, next to the attractive principle of universal freedom, the most potent among them is the rapid development of our railroad system. I do not refer alone to the direct influence upon those who come to find work in the actual construction of our roads, but to the indirect and far wider influence exerted upon those who come to make homes for themselves and their children upon our public domain, which railroads open for settlement, and bring within reach of markets.

It may be instructive to notice, for a moment, how immigration has kept pace with railway development in this country.

Decades.	No. of miles of railroad constructed.	No. of immigrants during same time.
From 1820 to 1830	41	244,490
" 1830 " 1840	2,156	552,000
" 1840 " 1850	5,289	1,558,300
" 1850 " 1860	21,295	2,707,624

Thus it will be seen that prior to the railroad era immigration was very small, and that during the last decade (the period of greatest railway development) immigration correspondingly increased. Since 1863, the average number has been about 300,000 per annum. It is not difficult to understand why, in a country like ours, this correspondence between the increase of railways and immigration should exist. Had it not been for the facilities they afford, the lands in the vicinity of water communication would have all been occupied, and those at a distance would have been worthless. The immigrant would have had none of the inducements to come here which are now presented to him. In fact, without railroads, Chicago and most of the other great cities of the West would be only second-class villages; and Minnesota, Iowa and Wisconsin would, to a great extent, be Indian hunting-grounds. It is because railroads annihilate the distance between cheap lands and good markets, and thus enable the farmer to dispose readily of his products, that the people of the Old World are so anxious to make homes among us. It is for that reason that among the people of my own State to-day, will be found more than a hundred thousand hardy, honest, industrious Germans, Norwegians and Swedes, who rank among the best farmers of the State.

Of course no mathematical estimate can be made of the money value of these immigrants to our country, but it is usually placed at about $1,000 per head. This estimate is based upon the assumption, that each free person adds as much to the wealth of the nation, as the former slave did to his master—an assumption evidently unjust to the free citizen. But in order that my calculation may be beyond the reach of cavil, I will estimate the money value of each immigrant at only $500, which will give an aggregate increase of national wealth from this source alone, of $750,000,000 within the last five years. It is also stated by John A. Kennedy, formerly superintendent of the Castle Garden immigration depot, in New York, that "a careful inquiry, extending over a period of seventeen months, gave an average of $100 (almost entirely in coin) as the money property of each man, woman, and child" landed at New York. At this estimate, the amount of gold and silver added to the circulation of this country every year, since 1863, is about $30,000,000, or the enormous sum of $150,000,000 in five years. Add this to the increase just mentioned, and we have $900,000,000 as the result of immigration during that time—a sum much greater than the entire estimated wealth of the United States in 1790. Dr. Engel, the Prussian statistician, says:

"Estimated in money, the Prussian State has lost during sixteen years, by an excess of 180,994 emigrants over immigrants, a sum of more than 180,000,000 thalers. It must be added, that those who are resolved to try their strength abroad are by no means our weakest elements; their continuous stream may be compared to a well-equipped army, which, leaving the country annually, is, after having crossed the frontier, lost to it forever." "A ship loaded with emigrants," says Dr. Engel, "is often looked upon as an object of compassion; it is, nevertheless, in a politico-economical point of view, generally more valuable than the richest cargo of gold dust."

What Prussia has thus lost we have gained. But why will the Northern Pacific railroad prove peculiarly valuable in stimulating immigration? First, because it will open to settlement the best portion of the public domain; second, because it will influence the best class of immigrants, by offering them the kind of lands they want, and in the climate they prefer. This human tide, which sets towards our shores, instinctively follows its own parallels of latitude. That part of Europe which lies north of the 48th degree of north latitude has furnished, probably, four-fifths of the foreign population now in the United States, and very much the greater part of that population came from Norway, Sweden, Prussia, and other parts of Northern Germany, all of which countries lie north of latitude 50°. An immigrant from Spain, Portugal, Turkey, Italy, or any of the countries south of the 48th degree would be a curiosity to the people of the Northwest, except that they may have occasionally made the acquaintance of some gentleman from the latter country, as he plied the enterprising avocation of "organ grinder." The intelligence, the energy, the thrift, the money, and the progress, all come from Northern Europe, and from a parallel of latitude further north than the line of this road. The census reports show, also, that but few of this class of immigrants seek the southern portions of the United States. *The construction of this road will open up a vast extent of fertile territory, which, in climate and natural productions, exactly meets the wants of the populations of northern Europe, and hence it willl exert a most powerful influence upon them.* Let the Northern Pacific railway aud the Homestead Law march hand in hand across the continent, and in less than a decade you will see along its line at least 2,000,000 of the hardy farmers, miners, and mechanics of Northern Europe, in whom alone the Government will find an increase of wealth, a hundredfold greater than the cost of the road; and who will bring with them enough gold and silver to pay five times over the amount of the proposed loan. If the United States are to advance in wealth in the future, as they have done in the past, we must keep wide open the doors to those people who bring with them, and who dig from our soil, so large a proportion of this wealth. And as no other road can, for the obvious reasons just named, supply the inducements to immigration which will be afforded by the Northern Pacific, it seems to me that this alone is conclusive as to the merits of its claim for aid.

2.—*It will be the means of developing our agricultural and mineral resources.*

We are sometimes told that " this is doubtless a meritorious and beneficent enterprise, but the Government cannot afford to aid it because of our immense national debt." We have already seen what other nations similarly situated have done, and how the results have demonstrated their wisdom. Let us now examine this objection by the light of the facts alone, unbiased by their example.

It is true our debt is large, but our resources are also immense. The true way to pay the former is to develop the latter.

The farmer who should refuse to incur the expense of plowing and sowing his fields because he is in debt, would be voted an idiot by all his neighbors. Wherein does he differ from the statesman, who, aware of the inexhaustible wealth of our soil, and of the countless treasures that crop out on almost every mountain-side, can yet refuse, on grounds of economy, (?) the perfectly safe loan of Government credit, to the comparatively small extent necessary to develop that wealth, and gather up and appropriate to the public benefit those treasures?

The public debt of a nation is great or small, according to the proportion it bears to the public wealth. A debt that would have crushed the United States in 1800 would not be felt to-day. Hence, in the exact proportion that our wealth increases our debt diminishes. For instance, in 1840, the entire national wealth was valued at $3,764,000,000. At the close of the rebellion, the national indebtedness reached over $3,300,000,000. Hence, to have paid the debt of 1865 twenty-eight years ago, would have required *ninety per cent. of all the* property, real, personal, and mixed, in the country. In September, 1868, our debt was $2,535,614,313; and, according to the most reliable estimates, the wealth of the nation had reached $30,000,000,000. *While, therefore, the debt of* 1865 *would have consumed almost the entire property, public and private, owned by the nation in* 1840, *the payment of the debt of* 1868 *would require only about eight per cent. of its present wealth.* It is therefore apparent that the burden of the debt of 1868, is only about one-twelfth as great on our present property, as the debt of 1865 would have been on the property of 1840. Prospecting the future by a just analogy with the past, let us see what will be the relative burden of our debt in the year 1900, on the assumption that the actual amount of the debt remains unchanged, and that the ratio of increase in wealth shall continue as great hereafter, as it was from 1850 to 1860. The ratio of increase during that decade was 126.45 per cent. I will assume, however, that the decennial increase will be only 100 per cent., and the following table will express the result:

YEAR.	National wealth.	Percentage of debt to property.
1860	$16,159,000,000	....
1870	32,318,000,000	7.84
1880	64,636,000,000	3.92
1890	129,272,000,000	1.96
1900	258,514,000,000	.98

For the data of this table I am indebted to Hon. D. A. Wells, who shows by a document published in 1864, that it is by no means an unreasonable estimate of our prospective wealth.

I think no one can reasonably doubt that our national capacity for growth in the future, is quite equal to what it has been, and that with wise statesmanship we may confidently expect to realize the increase just estimated. *It will, therefore, be seen that, in* 1900, *our debt will be only one-eighth as great a burden as it is now, or only about one-ninetieth what it would have been on the property of* 1840. *In* 1900, *a tax of only one per cent. will wipe out the entire indebtedness, while now it requires one per cent. to pay the current annual expenses of the Government.* Is it not, therefore, clear that the burden of debt diminishes, in exactly the same ratio, as the wealth of the nation increases?

Let us now inquire more particularly, how we shall best promote this increase of wealth, and thereby strengthen the public credit and reduce the public burdens. What has been the chief cause of this wonderful increase heretofore? Let facts and figures answer. A reference to the table of property already submitted, will show that the decennial increase of wealth from 1800 to 1840 averaged only about 36.7 per cent., while, for the next decade, it suddenly ran up to 89.6 per cent., and for the last one, from 1850 to 1860, it sprang forward to 126.45 per cent. The secret of this great increase in the ratio, may be found in the fact that the last two decades were the periods of railroad construction.

The relative growth in wealth, with the increase in railroads, is shown in the following statement:

TIME.	Miles constructed.	Increase of wealth.	Increase of wealth per mile of road.
From 1830 to 1840............	2,818	$1,111,000,000	$390,000
From 1840 to 1850............	6,203	3,371,780,000	540,000
From 1850 to 1860............	21,615	9,023,220,000	410,000

The entire accumulation of national wealth for two hundred years, from the settlement of the country down to 1830, (when the first railroad was opened,) was estimated at only $2,653,000,000. From 1830 to 1860 there were constructed 30,635 miles of railroad in the United States, and during that period our national wealth was increased to $16,159,000,000. *Thus it appears that during the thirty years of railroad development, there has been added to the nation's wealth, an amount more than six times as great, as the entire accumulations of the two hundred years prior to the inauguration of this wonderful engine of progress. In other words, the average annual increase of wealth prior to* 1830, *was only about* $13,265,000. *The average annual increase from* 1850 *to* 1860 (*the period of greatest railroad development*) *was* $902,322,000, *showing an average annual increase* 6,800 PER CENT. *greater during the last railroad decade than before* 1830. Such astounding results might seem to belong to the domain of fiction, were not their correctness demonstrated by the dry and incontrovertible logic of the census tables.

A little reflection will disclose the secret of this wonderful influence of railroads upon our national prosperity. The chief bases of wealth in this country are *lands, mines, labor.* Separated, they are each of little value. Combined, they produce the astonishing results we have seen. Railroads afford the only means of combining them successfully ; and hence to railroads must these results be mainly attributed. Without them, neither lands nor mines can be adequately supplied with labor, nor labor with a market. They are, therefore, the true philosopher's stone, which turns our mountains and prairies into gold.

In a very able work, entitled "Manual of Railroads of the United States," by Henry V. Poor, of New York city—a work, by the way, which should be in the hands of every railroad man and political economist in the country—I find this statement, viz : "The cost of transporting wheat for three hundred miles, over ordinary highways, will equal its average value at the point of consumption. Indian corn will bear transportation over earth roads only about one hundred miles." Hence all the land, outside of a radius of three hundred miles from the seaboard or river markets, would be, without railroads or canals, wholly worthless for the purposes of agriculture, and that which lies within such radius would depreciate in value, just in proportion as it approximated that verge of worthlessness. Take a map of the United States, and with scale and dividers mark out those little circles of trade which the wagon roads would accommodate, and then compare them with the present vast sweep of commerce on this continent, and you will be able to understand why the augmentation of national wealth bears so remarkable a relation to the increase of railroads. Strike from our great cities the support of those vast fertile empires, that lie outside of the earth-road radius; and from the prairies and rich valleys of the West take away the markets that are more than three hundred miles distant; and ruin, utter, hopeless ruin, will brood over almost every hearthstone in the land, and National Bankruptcy will stare grimly from the steps of the Treasury Department. The same causes which have thus strengthened our national credit, and augmented the public revenues, will continue, if properly encouraged, to produce similar results. Whether aided by the Government or not, the necessities of trade will eventually compel the construction of the road I advocate; but the cost is so great that private capital will hesitate to undertake it, and years will elapse before it will be done. It is only a question of time, it is true; but that question is an all-important one to the nation. It is a question whether we, who are now burdened with debt, shall take means to decrease that burden at once, and let the present generation reap the advantages that are so sure to result from the consummation of this great enterprise, or by a false economy refuse and postpone the incalculable benefits that court our acceptance.

Consider for a moment what will be the effect of the construction of this particular road, upon the credit and revenues of the country, by the development of the lands and mines in its vicinity. Take Montana as an illustration. So rich are the mines of that new Territory, that since 1862 the population has increased from a few hundreds to 60,000, and towns of from 4,000 to 10,000 inhabitants have sprung into existence. Its gold and silver products alone are estimated at $65,000,000 in six years. Observe under what disadvantages, and at what ruinous cost, this great addition has been made to the coin of the country. In 1867, 12,000 tons of freight were delivered in Montana from various distant points, on which the freight and insurance charges averaged 25 cents per pound, amounting in the aggregate to $6,000,000. This, in addition to the heavy expenses of travel to and from the Territory, and the loss of the use of capital for many months in the year by merchants and other business men, while goods are in transit, may enable us to form some conception of the burdens borne, for want of railroads, by those who open our mines and lay the foundations of States. The above statement of the cost of transportation is confirmed by Quartermaster General Meigs, who, under date of April, 1866, wrote: "From St. Louis to Helena, a town of 4,000 inhabitants, which has sprung into being in Montana within the past year, the estimated cost of transportation of freight by river and wagon trail is $500 to $600 per ton." And he adds: "How much must these enormous rates interfere with the opening and development of the rich mining, grazing, and agricultural regions of this Switzerland of America." Is it unreasonable to assume that with the reduced transportation and cheap labor this road will furnish, the gold and silver products of that Territory will be quadrupled, and that, instead of $12,000,000, as reported by J. Ross Browne for 1867, we shall have an annual production of at least $50,000,000?

The estimate for Idaho by the Government Commissioners was, for 1867, $6,500,000. This has been produced under disadvantages as great as in Montana. Her products would doubtless show as great an increase by the construction of the road, and the facilities it would afford. We may, therefore, I think, safely estimate the *increase* in gold and silver along the line of the road, which will result from its construction, at $50,000,000 per annum. Give to those people in Montana and Idaho (who, with freight at $600 per ton, have produced $20,000,000 of gold and silver per annum) reasonable facilities for the transportation of heavy mining machinery, cheapen labor and food by opening up easy access to the mining regions, and the next ten years will witness a development of mineral wealth, that the most sanguine among us cannot now anticipate. The safe and sure road to specie payment lies in this direction.

According to Redan, the average yearly production of gold and silver in all Europe is $29,024,000. We have seen that in the States and Territories on the line of this road, the production in 1867 was over $21,000,000. Are not its resources such as to justify the claim now made for it?

Let my own State illustrate the effect of railroads upon the increase of national wealth, by means of the facilities they afford for the settlement and cultivation of public lands. Without railroads to convey our products to the East, the rich lands of Minnesota would to-day be as valueless and tenantless, as are those which lie a thousand miles to the west of her. With them, see what has been accomplished! In 1861, Minnesota sent more soldiers into the field to fight for the Union, than she had white inhabitants in 1851. *In* 1857 *we imported our breadstuffs, and in* 1867 *we exported of wheat alone twelve million bushels, which sold at an average of about* $2 *per bushel, making in actual production of this one article, an increase of national wealth amounting to* $24,000,000. This, in addition to the amount consumed and retained for seed, and also in addition to the large exports of corn, beef, pork, potatoes, and other articles, which would amount to probably five or six millions more. President Polk, in his message of December, 1846, stated the value of all the exports of the United States at $102,141,893. Minnesota, then unknown, except as a supposed inhospitable, frigid, unexplored region, somewhere near the north pole, *in twenty-two years thereafter exported more than one-fourth as much as the entire national exports of* 1846.

The average annual exports of the United States for sixteen years, from the 30th of June, 1847, to June 30, 1863, were, of all kinds of agricultural products, (except cotton and tobacco,) $55,452,190, and of domestic products of every kind, $234,314,468.

These statistics, taken from the "Annual Record," disclose the astonishing fact that in 1867, *when only nine years old, the State of Minnesota exported of wheat alone, an amount equal to four-tenths, of the average annual value of all the agricultural products (excepting cotton and tobacco) of the nation, from* 1847 *to* 1863 ; *and an amount equal to one-tenth, of the average annual value of all the domestic exports of the United States, of every kind, including those of the sea, the forest, the mines, of agriculture, and of manufactures, during said sixteen years.* The total exports by the United States of domestic produce of all kinds, from 1800 to 1863, were $7,381,318,194, making an annual average for sixty-three years of $117,163,-780, or only about four times the value of Minnesota's wheat exportation for the year 1867. I make these comparisons with my own State because I am proud of her, and because it is by such comparisons we are able to form a correct estimate of the vast addition, the settlement and cultivation of our Western territories will make to the material wealth of the nation. The Governor of Minnesota estimates that only two per cent. of her lands have yet been brought under actual cultivation. Judge what she will do when one-half, or the whole, shall be cultivated!

The Northern Pacific road, by its intersections of the Red River, the Missouri, the Yellowstone, and the various other navigable streams on its line, will form a connecting link between the navigable waters within the boundary of the United States, at least a thousand miles in length; so that the combined length of the road and the river navigation with which it connects, will be not less than 3,000 miles. It will open to settlement the country on its own line and on said rivers, to the extent of from fifty to a hundred miles on both sides, making, within said limit of fifty miles, an aggregate of more than 150,000,000 acres. Deducting one half for uncultivatable lands (a very much larger deduction than the facts justify,) we have 75,000,000 acres of good, fertile land, such as that of my own State, that would be opened to settlement. Allowing that only one-third of the good land shall be actually tilled, we will have 25,000,000 acres under cultivation. The great results in Minnesota to which I have referred, were produced by the cultivation of only 2 per cent. of the whole, or about 1,000,000 acres. Multiply the products of Minnesota in 1867 by twenty-five, and we have 300,000,000 bushels of wheat per annum, which, at one dollar per bushel, will add $300,000,000 yearly to the nation's wealth; an amount six times greater, every year, than the proposed loan.

It will be observed that this estimate is only for the time immediately succeeding the construction of the road, and is by no means intended to represent even a tithe of what the productions of that country will amount to twenty years hence. It also excludes all that vast augmentation of wealth, which the road will effect, outside of the said narrow limit of fifty miles, and which will probably amount to very much more than I have estimated within such limit. Even in my own State, the value of every acre of land will be doubled by the construction of this great thoroughfare, and her present magnificent contributions to the national resources will be many times multiplied. Instead of being on the borders of civilization, she will then be in the very centre of that continent which is destined to be the centre of the world's commerce! When the main trunk line shall be completed, private enterprise will gridiron every part of that magnificent domain with branch roads, as it has done in the older States. Instead of 150,000,000 acres of land (now valueless) being brought within its influence, an area of *five hundred thousand square miles of fertile territory* will be transformed from the solitude, in which it has slept for centuries, into the teeming fields and prosperous homes of happy millions!

Consider also the value of the lands retained by the Government. Within the said fifty miles' limits, the Government would own over a hundred million acres. Estimating only one-half to be good, we have fifty millions of acres, now worth nothing, that would be made at once, by the construction of the road, worth from $1.25 to $10 per acre, making an aggregate addition from this source alone, of at least $100,000,000. The fact that the Government donates her lands to actual settlers does not affect this estimate, for the reason that such donation is made because the consideration of settlement and cultivation is clearly more profitable to the Government, than the money price the land would bring.

4. *It will augment our imports and exports, and thereby increase the public revenue.*

On this point, I ask the attention of the House to a few facts and figures taken from a paper read before the "Statistical Society of London," in November, 1866, by R. Dudley Baxter. He assumes it as an "axiom, that the commerce of a country increases in exact proportion to the improvement of its railway system, and that railway development is one of the most powerful and evident causes of the increase of commerce."

The following statistics are adduced, together with other important facts, in proof of that proposition. Any one who will give his paper a careful reading will, I think, be convinced that there is much truth in his theory.

Proportion of Exports and Imports to Railways and Navigation in Great Britain.

YEAR	Miles of railway and navigation.	Total exports and imports.	Exports and imports per mile.
1833	4,000	£85,500,000	£21,375
1840	5,200	119,000,000	22,884
1845	6,441	135,000,000	20,959
1850	10,773	171,800,000	16,006
1855	12,334	260,234,000	21,098
1860	14,433	375,052,000	52,935
1865	17,289	490,000,000	28,341

In 1833 there were about 4,000 miles of navigation, and exports and imports had been for some time stationary before that period.

Similar results are produced in France, as shown by the following table:

Proportion of Exports and Imports to Railways and Navigation in France.

YEAR.	Navigation (7,700 miles) and railways.	Exports and imports.	Exports and imports per mile open.
1840	8,264	£82,520,000	£9,985
1845	8,547	97,000,000	11,358
1850	9,507	102,204,000	10,750
1855	11,015	173,076,000	15,712
1860	13,286	232,192,000	17,476
1865	15,830	293,144,000	18,518

Here their is a steady rise in the amount per mile checked, only by the revolution of 1848.

Perhaps the most remarkable illustration of the relative increase of foreign commerce to railways may be found in Belgium.

Proportion of Exports and Imports to Railways and Navigation in Belgium.

Year.	Canals (910 m.) and railways open.	Exports and imports.	Exports and imports per mile.
1839	1,055	£15,680,000	£14,862
1845	1.205	26,920,000	22,340
1853	1,590	47.760,000	30,037
1860	1,907	72,120,000	37,818
1864	2,220	97,280,000	42.919

Compare Belgium with Holland. Holland was possessed of immense advantages in the perfection of her canals, which were the finest and most numerous in the world; in the large tonnage of her shipping; in her access by the Rhine to the heart of Germany; and in the command of the German trade, which was brought to her ships at Amsterdam and Rotterdam. The Dutch relied on these advantages and neglected railways. The Belgian Government adopted a wise and extensive system of railway improvements. Observe the results. In 1839 the Dutch exports and imports were £28,500,000, nearly double those of Belgium. In 1862 they were £59,000,000; when those of Belgium were £78,000,000. Thus, while Holland had doubled her commerce, Belgium had increased fivefold. The former possessed a territory 13,890 square miles in extent; the latter 11,313 miles. Their populations were not far from equal.

Various other statistics are given, proving conclusively that in all commercial nations, a distinct correspondence exists between the means of internal communication and foreign commerce; but I will content myself by referring to those of our own country.

Proportion of Exports and Imports to Railways and Canals in the United States.

Year.	Canals (6,000m.) and railways.	Total exports and imports.	Exports and imports per mile.
1830	6,041	£31,000,000	£5,130
1844	10,310	45,759,000	4,437
1850	13,475	68,758,000	5,102
1855	23,398	111,797,000	4,778
1860	34.770	158,810,000	4,567

In 1830 the imports and exports were	$155,000,000
In 1793 " " " " "	57,109,000
Increase in the 37 years preceding the railroad era	$97,891,000

Average annual increase, $2,646,243.
Average annual percentage increase, 4.63.

In 1861 the imports and exports were	$762,288,550
In 1830 " " " " "	155,000,000
Increase in the 31 years of railroad development	$607,288,550

Average annual increase, $19,589,950.
Average annual percentage increase, 12.57.

Showing an average annual increase, more than two and a half times greater since the beginning of the railroad era, than before.

But I must not extend these comparisons. Arrange the statistics of imports and exports in any way you choose; compare their increase with railroad development during any given period that may be selected, and similar results will be disclosed.

These statistics certainly show a very remarkable correspondence between the increase of railways, and the increase of imports and exports, and nowhere is it more apparent than in the United States. This increase will doubtless be much greater per mile, when we shall have completed the proposed Pacific roads, for the reason that they will develop an immensely rich mineral region, and be the great highways for the commerce of the world. The foreign commerce of a nation depends mainly upon the value of the products of its soil, its workshops, and its mines, and upon the facilities it affords for the internal interchange of those products, for the commodities of other countries. Hence, considering the vast extent of the United States, the millions of square miles wholly undeveloped, the sparseness of the population, and the difficulty in exchanging commodities between distant points, we may reasonably expect to find that the relation between the increase of imports and exports and railway development, will be more striking in this country than in any other. Since 1830, every additional mile of road has been followed by an additional twenty to twenty-five thousand dollars of foreign commerce. May we not, therefore, safely assume that the construction of the Northern Pacific, with all the advantages and probable results I have enumerated, will add to our imports and exports at least $20,000 per mile, or an aggregate of $40,000,000 per annum. As imports are generally about equal to exports, we shall have from this cause an additional $20,000,000 of imports, on which a tariff equal to the present, will yield an annual revenue of over $8,000,000.

THE NORTHERN PACIFIC ROAD IS A COMMERCIAL NECESSITY.

First, because one road will not be adequate to the business of the country; and, second, because no other route will so certainly control the commerce of Europe and Asia. I have conversed with some gentlemen on this subject, who seem to fear that there will not be room and business for more than one road. Let us briefly examine this objection. It is only 15 years, since by the opening of the *Cleveland* and *Toledo* railway, a continuous line was formed between Chicago and the eastern seaboard. Now there are no less than five double-track roads connecting that city with New York and Boston, and all of them are overwhelmed with business. The increase of the tonnage of these great through lines, furnishes one of the most remarkable illustrations of the wonderful material progress of the United States, as well as a conclusive answer to the objection under consideration. The following table, copied from the "Manual," will show the increase of tonnage on three of those lines for ten years:

Roads.	1857. Tons.	1867. Tons.	Increase. Percentage.
New York Central..........	838,791	1,667,926	98.85
Erie	978,069	3,484,546	256.26
Pennsylvania	530,420	4,000,538	654.24

Some half dozen or more roads connecting with these great lines, are pushing forward as rapidly as possible to form a connection with the Omaha and San Francisco road, which, it is supposed, is to do the business not only of the Pacific slope, but also of Europe and Asia. Now, sir, in view of the fact that fifteen years ago there was no railroad connection between Chicago and the East; and in view of the immense increase of internal commerce, that now crowds all the five great thoroughfares between those points, is it not absurd to suppose that one road can accommodate the trade, between the people of the Atlantic and Pacific slopes, and between the six hundred millions of Asia and the two hundred millions of Europe? The fact is, that the commerce of our own country will, in less than ten years, crowd with business two more Pacific roads. There is ample room for three. Mr. Baxter, from whose paper I have quoted, demonstrates that in Great Britain railroads will pay, that have three and a half miles of country on each side to support them. The question we debate is, whether a territory 1,500 miles in width: will not be crowded if we build three roads across it.

But it is in regard to its relations to foreign commerce, that the advantages of this road are most conspicuous. The most intelligent statesmen of Europe concede that the day is not far distant, when the trade between that continent and Asia, will be carried on across this country.

In 1864 a paper was read before the British North American Association, by Colonel SYNGE, of the Queen's Royal Engineers, in which he makes this statement: "America is geographically a connecting link between the continents of Europe and Asia, and not a monstrous barrier between them. It lies in the track of the nearest and best connection, and this fact needs only to be fully recognized to render it in practice what it unquestionably is in the essential points of distance and direction."

Says Lord BURY:

"Our trade (English) on the Pacific ocean with China and with India must ultimately be carried on through our North American possessions. At any rate, our political and commercial supremacy will have utterly departed from us if we neglect that just and important consideration, and if we fail to carry out to its fullest extent the physical advantages which the country offers to us, and which we have only to stretch out our hand to take advantage of."

English statesmen and capitalists are alive to the importance of this subject, and their Government has already sent out a corps of engineers, which has made a careful survey of the whole country from Canada to the Pacific, and reported that a railroad through the British North American possessions is entirely practicable. And in furtherance of this scheme, she has granted $20,000,000 to aid in the construction of the Grand Trunk road to Halifax. While, therefore, it is substantially settled that the commerce of Asia and Europe is to find its pathway across this continent, it is by no means conceded that that pathway is to be upon our soil, or that its incalculable benefits are to be enjoyed by our country. Great Britain, ever on the alert to increase her power and continue her commercial supremacy, can compete with us for this great carrying trade, by a railroad through her own possessions. That she will do so unless prevented by our superior diligence, no one who has studied her history can doubt. In view of these facts, and of the immense "physical advantages" the United States enjoy, important considerations arise in regard to the location of our Pacific railways.

It is now simply a question of statesmanship, whether the

United States or England shall monopolize and control the trans-continental commerce of the world, and whether the vast fertile country on our northern border shall become tributary to us, and eventually a part of our own territory, or remain forever under British dominion.

Commerce seeks the shortest and cheapest transit, and the nation that can supply these conditions will control it. We have seen that the northern route has the advantage of over six hundred miles, in its connections with the navigable waters of the Atlantic and Pacific, and that the distance from New York to the Pacific is four hundred and twenty miles less by this, than by the Central route. In addition to this important fact, it is also true that Puget Sound is some eight hundred miles nearer than San Francisco, to the principal Asiatic ports. This last statement is confirmed by all the well-informed persons from the Pacific coast with whom I have conversed on the subject, and the fact is accounted for by the prevailing ocean winds and currents. It is also confirmed by Hon. G. M. Dodge, chief engineer of the Union Pacific railroad. In a report made to that company, in December, 1867, urging the adoption of a line to connect the Union Pacific with Puget Sound, Mr. Dodge says:

"This inland sea includes the whole body of water flowing into the Pacific at the Straits of San Juan de Fuca, and surpasses the Mediterranean in the safety of its navigation and the security and anchorage of its harbors. *For all the China, Japan and India trade coming down the Pacific coast, it would save the water-transportation for the entire length of coast line from the Straits of San Juan de Fuca to San Francisco, as all ships coming from China, India, &c., make our coast near the mouth of these Straits.*"

Turning to the map of the world, we find the Amoor river taking its rise in Central Asia and flowing eastwardly to the Pacific, on a parallel of latitude some two or three degrees north of Puget Sound. This river is navigable for a distance of 2,600 miles into the interior of that continent. Its most considerable southern branch—the Soungaree—is also navigable to within a few hundred miles of Pekin. The Amoor is the Mississippi of the East, and through its waters we may have access by steam to the very heart of the Asiatic continent. Some conception may be formed of the commerce soon to spring up on that river, from the fact that at Kyachta and Mai-Mat-tschin, Chinese and Russian trading points, situated 400 miles west of its head of navigation, the imports and exports amount to $21,000,000 per annum. The Chinese trade has already assumed gigantic proportions. The exports of tea and silk from the single port of Shanghai for the year ending July 1, 1860, were as follows:

	Tea—lbs.	Silk—bales.
Great Britain	31,621,000	19,084
United States	18,299,000	1,554
Canada	1,172,000	....
France	....	47,000
Total	51,092,000	67,638

The Amoor river, but recently opened to commerce, is destined at an early day to become the chief outlet for the teas and silks of China, and for the skins, furs, and other commodities of Russia. Taking into the account the ocean winds and currents before referred to, and the fact that the mouth of this river is two or three degrees north of the latitude of the straits of San Juan De Fuca, it is evident that the western terminus of the northern road must necessarily be the point, at which the commerce of the country it drains will strike our shores. Beginning, therefore, at a point twenty-six hundred miles in the interior of Asia, and only a few hundred miles from Pekin, we will have by way of the Northern Pacific road almost an air line of steam communication to the principal ports of Europe, and only 1,775 miles of the entire distance will be by rail. Or if we estimate the navigable rivers on the line of the road, and having the same general direction with it, we may by the construction of an aggregate of only eight hundred and twenty-five miles of road, at different points, as already shown, furnish connected steam transportation between the interior of Asia and Liverpool. If it be true that commerce seeks the shortest and cheapest transit, surely this route is indicated as the great highway of the nations.

But if the English Government shall build the contemplated road through British America, it will have nearly all the advantages of shortness of line between the waters of the oceans, and also of proximity to the Asiatic continent, possessed by the Northern Pacific. If our Government shall neglect the claims of this road, and Great Britain shall construct hers, with the advantage over all others of more than a thousand miles in distance, aided, as it will be, by the commercial dominion of England on the seas, British America, and not the United States, will own and control the chief thoroughfare of the nations. The only road that can prevent the construction of the British line, or which can compete with it if built, will be the Northern. Its prompt construction will indefinitely postpone the other, and will in a short time, by means of emigration from this country to British America, and the affinities and necessities of trade between the two localities, peacefully, and without expense, annex that country to ours, where it belongs, and thereby secure forever to the United States the incalculable advantage of controlling the channels of the world's commerce. We will then be prepared to successfully challenge England's boasted commercial supremacy, and to gather the profits of our superior "physical advantages."

In addition to the many great pecuniary advantages, and benefits, to which I have referred, our Pacific railways may be made to furnish

A SINKING FUND FOR THE PAYMENT OF THE NATIONAL DEBT.

This proposition contains nothing new or chimerical. France, Spain, Portugal, Italy, Austria, and Holland, following the example of Belgium, have all adopted a system whereby their railways furnish a sinking fund for such purpose. A large portion of the public debt of all those nations will, by the application of the fund thus raised, be extinguished at the end of various terms of 85, 90, and 99 years.

In France it will, in less than 90 years, entirely relieve the people of the whole burden of the national debt of $2,206,000,000.

This sinking fund is raised, in the countries named, by the imposition of a small tax, or duty, on the earnings of their roads. Why may we not act upon this principle, and thereby make our railways the means of still greater relief to the tax-payers of the nation? The principle involved has been found practically successful in other countries; may it not be so in ours? Why shall not those great and profitable enterprises, which demand so largely the aid of the people, be made to reciprocate, by assisting directly, as well as indirectly, to bear the burdens of the people? This would seem to be especially proper and expedient, when it is borne in mind that these roads are to supply to Europe and Asia the shortest, quickest, and cheapest means of communication, and that, in consideration of the benefits thus to be conferred, it is but just that the people of those countries should pay a reasonable tribute.

It would, perhaps, be improper to attempt it upon roads built wholly by private enterprise and private capital; but not so in regard to the Pacific roads, which are so largely aided by the Government, and which are national in their character. They are the recipients of the nation's favors, and it is eminently proper that they should contribute directly, to the discharge of the national obligations.

A Government tax of only 2 per cent. on gross earnings, will prove no burden to the companies, and no injury to the business of the roads, but if properly applied as a sinking fund for the payment of the public debt, it will in a few years produce astonishing results. I suggest *gross* earnings instead of *net*, for the reason that there will be less opportunity for fraud in making returns.

The average gross earnings of the "*New York Central,*" "*Pennsylvania,*" "*Erie,*" *and* "*Michigan Central*" roads for

1867 were $26,200 per mile. The aggregate length of their trunk lines to the Pacific will be, in round numbers, 6,000 miles. Their gross earnings, estimated at the same rate with the four roads just named, will amount to an aggregate of $157,200,000 per annum. This is, doubtless, far below what such earnings will be a few years hence; but, for the purpose of illustrating my proposition, I will estimate them at only $100,000,000 per annum, on which a Government tax of 2 per cent. will yield an annual sinking fund of $2,000,000. This fund, if invested every year for 90 years, at 4 per cent. compound interest, will, at the end of that time, amount to $2,560,812,000, a sum exceeding the entire debt.

If this sinking fund of $2,000,000 per annum be applied annually to the purchase of United States bonds, drawing 4 per cent. interest, it will, in ninety years, (if the interest accruing on all bonds purchased be annually re-invested in bonds at the same rate,) result in placing in possession of the Government every dollar of its bonds, and in relieving the people of the entire burden of the debt; provided, only, that other means shall be used for the payment of the interest.

In other words, if a sufficient sum be raised by taxation, or otherwise, to pay the interest on the debt, this railway sinking fund will, (if it be annually invested, and re-invested, in the manner proposed,) in ninety years, absorb the entire principle of the debt.

If, instead of being used as a sinking fund as just proposed, it be applied when received, to the redemption of the bonds, it, of course, will not effect the entire extinguishment of the debt within the period named, but the relief it will afford to the people by the retirement of the bonds, and the consequent reduction of annual interest, will in the aggregate amount to the same sum.

Now sir, I would authorize the funding of our entire bonded debt into American consols, having ninety years to run, drawing 4 per cent. interest, principal and interest expressly payable in specie. I would then, by a temporary loan of Government credit, properly secured, aid the construction of at least two more roads to the Pacific, and impose upon them a tax such as suggested. The money raised by this tax should constitute a sinking fund, to be applied to the purchase of the 4 per cent. consols, in the manner described; the result of which would be an entire absorption of the debt in ninety years. Certain and adequate provision being thus made, for the accumulation of a fund sufficient to pay the debt, capitalists at home and abroad would readily invest in so desirable a security. While bearing a higher rate of interest than the debts of other nations, the security would be equal to theirs, and the duration of the consols would commend them to those who desire to invest on long time. In this way, our entire bonded debt may be speedily funded at 4 per cent. interest, which reduction of interest, will at once diminish the current expenses of the Government more than $42,000,000 per annum. *The effect, therefore, of this proposition, (if adopted,) will be to reduce the taxation of the people to the extent of* $42,000,000 *per annum; and at the same time provide a sinking fund that will ultimately relieve them of the entire burden of the debt.*

The fact, therefore, that we have an immense debt which these roads will so largely assist in paying, is one of the strongest arguments in favor of the aid proposed. We have already seen that the burden of the debt decreases, in exactly the same ratio, that the national wealth increases. The relation of the debt to our present property, is as one dollar to twelve dollars. Double the amount of national wealth, and it will be as fifty cents of debt, to twelve dollars of property. In other words, the increase of property *one hundred per cent.* is equivalent to the decrease of the burden of the debt *fifty per cent.*—that is to say, the addition of $1,200,000,000 to the wealth of the nation, is equivalent to the actual payment of fifty millions on the debt. In the same manner, and in the same ratio, the increase of wealth will diminish the burden of annual interest on the debt. Hence, while by the construction of these roads, the national wealth will be increased many hundreds of millions, and the burden of annual interest will be thereby proportionately diminished, they may also be made to furnish a sinking fund, which, properly applied, will greatly reduce the *rate* of interest, and finally pay the debt itself.

Of course this system cannot be adopted unless the tax to be imposed shall be the same on all the Pacific roads, including those already aided by the Government. To impose it upon one, and not upon all, would be to make a distinction that would work injuriously. There will be, I apprehend, no practical difficulty on this point.

In calculating the period within which the proposed sinking fund will extinguish the debt, I have assumed that the interest on our bonded debt is to be reduced to four per cent. If the rate should be higher, the time required for such absorption would be proportionately *shorter*.

I have assumed that the probable gross earnings of the Pacific roads, will be, *for the next ninety years*, $10,000 per mile less than the present average aggregate earnings of the *N. Y. Central*, *Erie*, *Pennsylvania*, and *Michigan Central* roads. A moment's consideration will show that this estimate is unreasonably low. Forty-one years ago the entire tonnage crossing the Alleghany mountains, both ways, did not exceed 30,000,000 pounds per annum. Thirty-two years ago, the annual tonnage coming from the West, through the Erie canal, amounted to only 112,000,000 pounds. The Commissioner of the General Land Office states, in his recent able report, that "wagon freights to the Pacific are now estimated at 230,000 tons (460,000,000 pounds) per annum"—an amount *fifteen times greater* than the tonnage crossing the Alleghany mountains in 1827, and more than *four hundred per cent.* in excess of the tonnage from the West, through the Erie canal, in 1836. Within the last ten years, the business of all the railroads in the United States has increased from 100 to 600 per cent. At this rate of increase, how shall we estimate the probable earnings of these great interoceanic thoroughfares, when, thirty years hence, the trans-continental and local business of a nation numbering *one hundred million people*, and owning *two hundred and fifty billion dollars of property*, shall be poured upon them? And to all this must be added the trade of China, Japan, India, and Europe—"a trade along whose slow and painful track, when it was conducted by beasts of burden, and by oars, and sails, instead of by the iron-horse and ocean steamship, great cities sprung up in the desert sands of Asia and on the coasts of the Mediterranean. Babylon, Nineveh, Palmyra, Bagdad, Damascus, Constantinople, Alexandria, Rome, Venice, Genoa, and London, are the outgrowth of this trade in former centuries." A trade that will hereafter "found such cities in the new, healthful, and inviting regions through which its flow is destined to enrich the world; and Oregon (and Washington) as well as California, Montana as well as Utah, will hereafter have their San Franciscos, Chicagos, St. Louises, Cincinnatis, and New Yorks—great emporia of an internal commerce heretofore unknown, as well as of the world-encircling commerce of the Indies."

Some conception of the extent of this traffic, may be formed from the fact that even now, while it finds its tedious way around the Cape of Good Hope, across the isthmus of Suez, or by weary and painful caravan routes across the Eastern continent, its estimated amount is $300,000,000 per annum; and also by the fact stated in the "Parliamentary Papers" of 1866, volume 26, that the importations of *tea alone* into Great Britain, from China and Japan, for three years, were, for 1863, 136,803,218 pounds; for 1864, 124,359,243 pounds; for 1865, 121,156,712 pounds. Total, 382,319,173 pounds.

In view of all these facts, I submit that a tax of two per cent. on the gross earnings of these roads, will, long before the expiration of the present century, be far more likely to exceed *six million dollars* per annum, than to fall below *two millions;* and hence the payment of the debt may be, in this way, effected much earlier than the time named.

I have said that the Northern route seems to have been pointed out by the hand of nature as the great highway of the nations. In the words of Gov. Fuller, "the road, as projected, follows the wavy outline of the isothermal temperate zone of the northern hemisphere, along which civilization makes the circuit of the globe. This zone contains the zodiac of empires. Along its axis appear the great cities of the world. Along this belt, encircling the globe, we discover the chief centre of intelligence and power, and the greatest intensity of energy and progress. Along this axis, hardly varying one hundred leagues, civilization has travelled as by an inevitable instinct since creation's dawn." In its topographical and meteorological conditions, also, this route appears peculiarly adapted to the great purpose for which it was evidently designed. Oceans, lakes, and rivers seem to be located for its advantage. The rough and rocky summits of the mountains are depressed to afford it a convenient passage. Warm breezes from the Southern ocean meet the tempered winds from the east in its mountain passes, and diffuse over them a mild and healthful climate. Ocean currents and winds sweeping over the Pacific from the southwest, force the commerce of Asia to its western depot, while one of the finest inland seas of the world affords that commerce a safe and convenient anchorage.

Now, sir, in addition to all these physical advantages, we have seen that this enterprise will require no outlay of money by the Government, but, on the other hand, it will diminish the public expenditures, and, by the development of our immense resources, vastly enlarge the basis of public credit, augment the national revenues, and thereby reduce the public burdens. The people understand these things, and hence their anxiety that Congress shall prove equal to the high statesmanship which can discern the great future before us. State legislatures have memorialized us to aid it. Boards of trade in nearly all the principal cities have endorsed it. And but a few days ago the National Board of Trade, which met at Cincinnati, representing the opinions and wishes of the shrewd, conservative merchants, manufacturers, and other business men and tax-payers of nearly every city in the Union, resolved, with great unanimity, "that this National Convention earnestly urges upon Congress the patriotic duty of granting immediate and adequate aid to perfect our system of Pacific railroads," which "system," the resolution says, requires the construction of both the Northern and Southern lines. It is demanded by the whole people: by the merchant, because it will enlarge the area of his trade; by the manufacturer, because it will increase the demand for his products; by the farmer, because it will afford him a market; by the miner, because it will enable him to command the appliances necessary to success; by the laborer, because it will enlarge the field of labor, enhance the value of its products, and necessarily increase its compensation; by the enlightened statesman and true political economist, because it will augment the public wealth and diminish the public burdens; by the philanthropist, because it will afford a peaceful solution of our Indian troubles; by the patriot, because he discerns in it, and in kindred enterprises, the true secret of national prosperity and progress.

Gentlemen may imagine that a pompous pretence of economy, will blind the eyes of their constituents, and increase their own temporary popularity; but let them be assured that they who presume upon the ignorance of the people, will find themselves sadly disappointed. The people of this Republic are proud of her. They know her resources and are determined to develop them. They foresee the grandeur and glory to which she may attain, and resolve to realize them. Development is the watchword of the century! In the world's great onward march, America must keep step to the drum-beat of the nations, or fall behind, a disgraced straggler from the ranks. We may, by our narrow views and mistaken ideas of economy, retard, but we cannot wholly obstruct the nation's progress.

By that economy which we all profess; by that progress, of which we are all proud; and in the name of the people whom we serve, I appeal to gentlemen to give this enterprise their support.

SPECIAL REPORT

OF

A RECONNOISSANCE OF THE ROUTE

FOR THE

NORTHERN PACIFIC RAILROAD

BETWEEN

LAKE SUPERIOR AND PUGET SOUND,

VIA THE COLUMBIA RIVER,

Made in 1869,

BY W. MILNOR ROBERTS, U. S. CIVIL ENGINEER,

UNDER THE DIRECTION OF

MESSRS. JAY COOKE & CO.

BANKERS.

ORGANIZATION.

Trustees for the First Mortgage Bondholders:

JAY COOKE,
J. EDGAR THOMSON.

Officers of the Northern Pacific Railroad Company:

J. GREGORY SMITH, *President*,	120 BROADWAY, NEW YORK.
R. D. RICE, *Vice-President*,	
SAMUEL WILKESON, *Secretary*,	
A. H. BARNEY, *Treasurer*,	
W. MILNOR ROBERTS, *Engineer-in-Chief*,	
EDWIN F. JOHNSON, *Consulting Engineer*,	

Board of Directors:

J. GREGORY SMITH,	ST. ALBANS, VT.
R. D. RICE,	AUGUSTA, MAINE.
THOMAS H. CANFIELD,	BURLINGTON, VT.
WM. B. OGDEN,	CHICAGO, ILL.
WM. G. MOORHEAD,	PHILADELPHIA, PENN.
WM. G. FARGO,	BUFFALO, N. Y.
B. P. CHENEY,	BOSTON, MASS.
GEO. W. CASS,	PITTSBURG, PENN.
FREDERICK BILLINGS,	WOODSTOCK, VT.
WILLIAM WINDOM,	WINONA, MINN.
JAMES STINSON,	CHICAGO, ILL.
SAMUEL M. FELTON,	PHILADELPHIA, PENN.
CHARLES B. WRIGHT,	PHILADELPHIA, PENN

Secretary: SAMUEL WILKESON.

Executive Committee:

J. GREGORY SMITH,
R. D. RICE,
WILLIAM B. OGDEN,
GEORGE W. CASS,
WILLIAM G. FARGO,
WILLIAM WINDOM,
S. M. FELTON,
CHARLES B. WRIGHT.

Financial Agents for the Railroad Company:

JAY COOKE & CO.
PHILADELPHIA.

Philadelphia, June 1, 1869.

W. MILNOR ROBERTS, Esq.

AND OTHERS:

GENTLEMEN:—The Directors of the Northern Pacific Railroad, having proffered us the Financial Agency of their Company, we feel it our duty before accepting the trust, and before we institute measures for the sale of the Securities of the Company, or become identified with this great work, to cause a thorough examination of the route to be made, by agents chosen and sent out by ourselves; and we do this, not because we do not place implicit confidence in the representations made to us, or doubt in the least the numerous and very able explorations already made in detail by men of the highest scientific and personal character, but because of our long established rule to make "assurance double sure," and to take every precaution to avoid placing in jeopardy, not only our own means, but the means of those who confide in our judgment.

We request you and the gentlemen detailed as your companions, to proceed to the Pacific coast and, after a thorough examination of Puget Sound and the Columbia river—the two termini of the Northern Pacific Road—to proceed eastward along the general line of the road, via Columbia river or the Snoqualme Pass, by such route as you shall select to the passes in the Rocky Mountains, and thence to Fort Benton, and also to the waters of the Yellow Stone.

Other parties, under Governor Smith, of Vermont, and Governor Marshall, of Minnesota, will explore the already well known route from Lake Superior to Red river, and up the valley of the Cheyenne to the great Bend of the Missouri. General Hancock, the military commander of the Northwest, now about returning from an extensive tour along the Missouri and the Yellow Stone, will furnish the information we need as to this remaining link.

As soon as possible after your return we desire a full Report from you of the whole route, based upon the explorations of these parties, as well as the information given at an earlier day by the distinguished explorers who have gone before you, and which we doubt not will now be fully verified.

Sincerely your friends,

JAY COOKE & CO.

To

Messrs. JAY COOKE & CO.

BANKERS,

PHILADELPHIA.

September 25th, 1869.

GENTLEMEN:—Pursuant to your instructions, dated June 1st, 1869, I proceeded, without delay, by railroad to Omaha, and there met the gentlemen, five in number, with whom I was to be associated in the important investigation committed to us. We proceeded by the Union Pacific and Central Pacific Railroads, 1,775 miles to Sacramento, and thence 120 miles by the Sacramento river and across the bay to San Francisco. At San Francisco, on ascertaining that the steamer for Portland, Oregon, would not sail for several days, we decided on taking the overland route back through Sacramento, and thence along nearly parallel with the Pacific coast to Portland, on the Willamette river, an important tributary of the Columbia. Railroads are already in the course of construction, extending northward from Sacramento, and we rode over the finished portion, extending 70 miles beyond Sacramento, to Oroville; thence in stages 560 miles to Portland, passing through a number of towns and cultivated lands in Northern California, and through Salem and other towns in Southern Oregon. Although this coast route is partially broken by some mountain ranges, there are extensive areas of fine agricultural country, which, on the completion of the railroads now in progress, will contribute their share of passenger travel and general traffic to the main trunk roads with which they connect. At the northern end of our journey, especially on the last hundred miles of the Willamette valley, approaching Portland, we were particularly struck with its beautiful appearance as an agricultural region, already settled and cultivated far beyond our anticipations. Two railroads, one on each side of this great valley, are now in process of construction, and the people of that region, with good reason, are confident that there will be business enough for both as rapidly as they can be extended. This splendid valley, with the aid of these railroads, will concentrate a large railroad trade at Portland, one of the western termini of the North Pacific Road, which must tend to augment the business and population of that flourishing city, and thus furnish a valuable contribution to the business of the Northern Pacific Railroad. The enterprising men of Portland, who, upon the occasion of our visit, gave us a hearty reception, and expressed the most lively interest in the success of the Northern Pacific Railroad, have, in the space of a few years, built an elegant city, with all the modern appliances of water supply, gas, Nicholson pavements, handsome churches, stores, etc., as well as commodious wharves, at which we saw numbers of sea-going vessels from foreign ports, and river and ocean steamers. Vessels drawing fourteen feet water can enter and depart at all times, and sometimes vessels of greater draft arrive. The population is already over eight thousand. One circumstance in connection with the enterprise and business capacity of Portland is worthy of special notice. The Oregon Steam Navigation Company, originated here by a few gentlemen less than ten years ago, with a capital of about one hundred and fifty thousand dollars, now owns twenty steamers running on the Willamette river, up and down the Columbia river from Portland, on Puget Sound, on two different stretches of the Columbia above the Cascades, on Lake Pend d'Oreille, and on two different portions of Clarke's Fork of the Columbia; they own two portage railroads, of their own construction, one six and the other fourteen miles long, and their capital is now over two millions of dollars, besides paying large dividends; so that the Company has become, deservedly, a power on this part of the Pacific slope.

To the liberality and kindness of Captain I. H. Ainsworth, President, and Mr. Reed, Treasurer of this important organization, our party are under very great obligations for the free use of their steamers, wherever we could avail ourselves of them to facilitate our examinations; on Puget Sound, on the rivers, and on Lake Pend d'Oreille. In no other way could we have made some of our most important observations so satisfactorily, and in the shortest possible time. There is, perhaps, no single fact to which I can refer in connection with our investigations more suggestive of the rapid and substantial growth of trade and commerce on this part of the Pacific coast, than the eminent success of this Steam Navigation Company, and the flourishing condition of Portland, the home of its far-seeing and public spirited proprietor, Captain Ainsworth.

Leaving Portland on one of the Company's admirable steamers, we steamed down the Willamette river a few miles to the Columbia river, and down that noble stream to the mouth of the Cowlitz, and up the Cowlitz a short distance to the town of Monticello, which is sixty miles by water from Portland. Here we disembarked, and were conveyed in coaches, partly along the valley of the Cowlitz, and thence over the portage between the waters of the Columbia and those of Puget Sound, to *Olympia*, a flourishing town at the very head of Puget Sound, a distance of eighty-five miles from the mouth of the Cowlitz river, and one hundred and forty-five, by our travelled route from Portland. The country between the mouth of the Cowlitz and Olympia, on the stage route, is very hilly, part of the way, but the valleys on either side of the portage mentioned, present a remarkably favorable route for a railroad with easy grades between the valley of the Columbia and Puget sound.

At Olympia, Captain Ainsworth, who kindly accompanied us, placed another steamer at the command of our party, and with her we made the tour of the Sound, stopping at the principal ports and noting all objects of interest. The citizens of Olympia take a warm interest in the Railroad, and extended to us numerous courtesies, and gave us much information. Olympia is in latitude 47°.03 north; longitude 122°.55 west. The magnetic variation on this part of the coast is about $20\frac{3}{4}$° east.

From Olympia we coasted along the eastern shore of the Sound, visiting *Tacomah,* about thirty miles from Olympia, in Commencement bay, where a large new saw-mill was just going into operation. It has six steam engines, and two hundred and fifty tons of machinery. There is deep water here, sufficient for the largest vessels. The tide-rise is usually about thirteen feet, but sometimes it is sixteen feet or more.

Tacomah is situated at the mouth of the Puyallup river. An Indian Agency is still kept up on the left bank, near the mouth of the river. It may be stated here that, with the exception of Olympia, all the ports on the Sound have at their piers an abundant depth of water for the largest sea-going vessels. Olympia harbor, at its upper end, is only navigable for large vessels when the tide is in; the shoal water extending about a mile and a-half from the piers. In approaching Seattle we passed Point Alki, and about two miles farther, just before entering the bay, Lamb's Head. The Freeport Mills are just around the point of Lamb's Head, and the Territorial University of Washington Territory stands on an eminence to the left of the town. The citizens fired a salute in honor of our arrival, and expressed the liveliest interest in behalf of the Railroad; hoping that Seattle would be selected as one of its terminal points. The land rises rather abruptly immediately back of the town, to the height of about three hundred feet above the sea, and then gradually falls and rises in heavy undulations, till at three miles back it falls to the level of Lake *Washington,* a fresh water lake, which is said to be only about twenty feet above high tide. The lake is between twenty and twenty-five miles long, and about eight miles wide. It has a large island near the middle. At the northern end of Lake Washington there is an outlet to a smaller one, called Lake *Union,* which is said to be about seven feet above high tide. Lake Union has an outlet to the Sound, near Seattle. There is also an outlet from the southern end of Lake Washington, by which the water flows into the Duwamish river, which runs to the Sound south of Seattle. The water of Lake Washington is perfectly clear and pure, and it is quite deep. There are here combined, water and water power for supplying a city. We rode over to the lake, and went in a boat obliquely across to its eastern side, about nine miles, and walked two and a-half miles over undulating ground to examine a coal mine, which had been opened but a short time. We found a regular vein of bituminous coal, of four to five feet thickness—a very fair article of coal for blacksmith and other purposes. Afterward, in an interview with Mr. L. B. Anderson, the pioneer of coal discovery along the Sound, I learned that he had traced this vein and others for six miles east and west, and that the lowest and best vein was sixteen feet thick—next to the basaltic rock. At Seattle the heaviest ships load with lumber for foreign ports. In going out to the north from Seattle bay, we pass West Point. At the head of the bay of Seattle is the Duwamish river.

From Seattle we steamed over to *Port Gamble* or *Tee Ka Let*, where we found important saw-mills, belonging to Pope and Talbot—the largest now on the Sound. These mills have made one hundred and fifty thousand feet board measure in twenty-four hours. When their new gang saws, now nearly finished, are complete, their capacity will be two hundred thousand feet per day. This firm owns and sails seventeen vessels to San Francisco, the Sandwich Islands, etc. They have had eight vessels at a time loading lumber at their wharves. They employ a steam-tug for towing large rafts. I present these statistics, in order to convey an idea of the magnitude and importance of the lumber business of Puget Sound. Without describing each locality where there are saw-mills, I may state that there are ten establishments now running, whose average daily product is from five hundred to seven hundred thousand feet, and the number and capacity of the mills is on the increase, and the business is yet in its infancy.

From Port Gamble we went around to *Port Ludlow*, which is also on the west side of the Sound, or rather of Admiralty Inlet, in about latitude 47° 54″. On entering the bay we passed around Port Tala. Here is found another extensive saw-mill, at Ludlow's Landing, cutting sixty-five thousand feet per day. There were two ships loading there at the time of our visit. Thence we went on to *Port Townshend.* (Point Hudson is in latitude 48° 07′ 03″.) This is a beautiful and perfectly protected harbor, and the town, built partly on the hill and partly under the hill, presents a very pleasing aspect. The island opposite the town across the bay is famous for abundance of deer. Port Townshend is now the first United States port of entry after leaving the British Possessions, or on entering by the Strait of Juan de Fuca from any foreign port.

From Port Townshend we steamed obliquely across in a direction a little north of east to Bellingham bay, on the eastern side of the Sound. Bellingham bay is three miles wide and fourteen miles long, extending from latitude 48°.33′ to latitude 48°.48′, with a depth of water ranging from three to twenty fathoms, with good sticky bottom for anchorage. Bituminous coal mines have been worked at *Sehome* and Whatcom, on this bay, for a number of years, and, recently, some new mines have been opened. There is no doubt that coal exists over an extensive area of country east of Puget sound; and as the demand on the coast is steadily increasing, it will be developed to meet the wants of commerce.

In passing the upper part of the Gulf of Georgia we were presented with a charming view of that noble body of water. Indeed, the entire voyage around the Sound, through Admiralty Inlet, the Straits of Juan de Fuca, the Gulf of Georgia, and Bellingham bay, including the views of various prominent points, lovely islands, and distant mountains, was but a passage through a continuous panorama of splendid scenery, which can scarcely be excelled in any part of the globe.

On our return from Bellingham bay, where we were within a few miles of latitude 49°, the extreme northern boundary of the United States, we visited *Victoria,* on Vancouver's Island, a handsome place, situated on a point on a small bay. It was much more flourishing a few years ago, during the Frazier river excitement, than it is now. At present, business is dull, and the local taxes are very high. Bituminous coal is sold at Victoria for $4.00 per ton; and anthracite coal is found on Queen Charlotte's Sound, and sold in Victoria.

From Victoria we steered across the Strait de Fuca, obtaining a view of Race Rock Light, also of the remarkable spit projecting six miles into the strait, on which stands the new Dungennes Light House. Thence passing over in sight of Washington Harbor, Protection Island, and back to Port Townshend, where we had to report ourselves as coming from a foreign land. We then returned, touching at the different ports, to Olympia; having made a complete circuit of this remarkable body of water, in which we saw and noted much more than could be conveniently embodied in this preliminary Report. We are under obligations for the great liberality and kindness of Captain Ainsworth, and all the officers of the steamers. It also affords me pleasure to thank Governor Evans, and Captain James S. Lawson, of Olympia, for much valuable information. Captain Lawson, who is in the service of the government, has spent years in the survey and study of these waters. He accompanied us on our trip, and described all important points of interest in a very satisfactory manner. We could not have made an examination of the channels and harbors on these glorious waters under more favorable auspices; and with the additional aid of the coast survey reports and charts, I have been enabled to obtain a very accurate knowledge of this important portion of our Pacific possessions.

I have likewise devoted considerable attention to the reports and charts of the mouth of the Columbia river, and the adjacent coast on both sides. In regard to the Columbia river, at and near its mouth, it may only be necessary to say in this place, in general terms, that it is entirely practicable to construct a Railroad with easy grades and curvature, at moderate cost, from the Willamette river down to any place that might be selected as a commercial terminus on the lower end of the river. Whatever difficulty sailing vessels may experience sometimes in entering the mouth of the Columbia, on account of adverse winds, the San

Francisco steamers experience none in making their regular trips between San Francisco and Portland, passing between the Ocean and the Willamette river, via the Columbia river.

It would consume too much space in this Report to present in detail all the knowledge acquired respecting the waters of Puget Sound, and of the lower Columbia river, in connection with proposed terminating points on the Northern Pacific Railroad. Without at present indicating any particular port as likely to be the best and most advantageous, it may be said that there is a choice of good sites for a large commercial city, such as must soon grow up at the terminal point of the Northern Pacific Railroad on the Pacific coast.

PUGET SOUND.

In referring to Puget Sound, in a general way, the mind naturally embraces the entire expanse of waters of which, technically, Puget Sound, the extreme southern portion, constitutes but a small part. The Gulf of Georgia, and the Strait of Juan de Fuca, are the two grand inlets from the Ocean; each being abont ten miles wide in their narrowest parts, and generally much wider. The Strait of Juan de Fuca must always be the main commercial avenue from the sea to the extensive region surrounding the vast interior salt water navigation, which permeates that portion of United States territory lying between latitude 47° and 49°, and between longitude 122° and 124¾°. The coast line of all these inland seas covers a distance of 1,800 miles, surrounded on the eastern side by magnificent forésts of pine, fir, cedar, etc., surpassing any forests elsewhere to be found on the globe, in the quantity and quality of the timber. Numerous settlements already exist at different points, generally where saw-mills could be conveniently located, adjacent to the valuable timber tracts, and with their piers so arranged that the largest Ocean ships can lie there in perfect safety, at all times, and receive their cargoes of lumber directly from the mills. The facilities for the greatest lumber trade the world has ever known, are here; and before the North Pacific Railroad can be completed, the product of the mills around Puget Sound will unquestionably exceed a million of feet per day, or three hundred millions per annum. With the settlements on these shores, which will accompany and follow the construction of this Railroad, and with the interior demand that it will create along the line of the road itself, it is safe to claim that on these inland seas will spring up, almost like magic, a trade which will establish this as the world's chief lumber mart.

There is not anywhere else on the globe to be found an unoccupied field for the establishment and permanent support of a new great city, such as should form the terminus of a Continental Railroad, uniting the waters of the Pacific and Atlantic by the shortest line between the great Puget Sound indentation of the coast in the west, and the Lake Superior indentation of the coast on the east. Between these extreme points the distance by a direct line is only about 1,350 miles; being thirty degrees of longitude of forty-five miles to each degree, between the latitude of 46° and 48°.

Next to the lumber trade in importance will be at first tne great fisheries off the Pacific coast; the facilities for the accommodation of which will be found at the terminus of this Railroad, where the vessels will be built, equipped, and found, complete, with every needed appliance, and manned ready for sea. I say "at first," because the day is not far distant when the manufactures which will grow up around this world of waters, will engender a commerce far exceeding that arising from the fisheries; and, as the forests recede under the insatiable demands of an increasing growth of population, agricultural products will fill the apparent void, for it is certain that the soil where these vast forests now grow is remarkably prolific. And if at some period in the future when numerous flourishing cities shall have grown up with the growth of this Pacific coast, the timber should be exhausted, a bountiful Providence has stored up for the use of the coming generations an abundant supply of coal, an article which is the basis of most of the wealth of Great Britain, and which, more than any single product of tho mines, has enabled the United States to take her present stand among the nations.

The climate of this favored region is very remarkable, and will always remain an attractive feature; and which must, therefore, aid greatly in the speedy settlement of this portion of the Pacific coast. Even in the coldest winters there is, practically, no obstruction to navigation from ice; vessels can enter and depart at all times; and the winters are so mild that summer flowers, which in the latitude of Philadelphia on the Atlantic coast, we are obliged to place in the hot-house, are left out in the open garden without being injured. The cause of this mildness is usually, and I think correctly ascribed to the warm water equatorial current, which, impinging against the Pacific coast, north of the Strait de Fuca, passes along nearly parallel with the shore, diffusing its genial warmth over the land, far into the interior. Of the fact there is no doubt, whatever may be the cause.

The opening of the Union Pacific and Central Pacific Railroads, forming part of the continuous railroad communication between the Atlantic and Pacific, is the signal for the commencement of the work which will open, at the earliest practicable moment, the direct line of Railroad by the valley of the Columbia to the Pacific tide water on the river and on Puget Sound. In my opinion this has settled the whole question; for it is only necessary that intelligent men should visit this portion of the Pacific coast to become impressed, as I have been, with its wonderful advantages. Until the opening of that continental line of Railroad, only a few, comparatively, would voluntarily undergo the toil and privations of a laborious journey of five or six months; and few, comparatively, could afford to pay the cost by the ocean and isthmus route; consequently the settlement of the Pacific slope has been slow, in past years, compared with what it will be when cheap fares and quick transportation shall enable hardy and industrious emigrants to pass over on the continental rail. I am not among those who have been so ready to exclaim that the Union Pacific and Central Pacific lines will prove to be at first comparatively unremunerative. They ought not to be when success is certainly attainable by proper management; the key note of which is, low fares for passengers and freights.

The precise point for the ocean terminus of the North Pacific Railroad cannot well be determined till the final route for the main line shall be adopted; but the foregoing remarks are applicable in any event.

I have dwelt longer than I anticipated upon this end of the route, but the importance of the subject is such that it cannot be satisfactorily disposed of in a few words.

Our party returned from Olympia and passed over to the Cowlitz river, where we embarked in a large canoe managed by Indians, who conveyed us safely down the river, a distance of about thirty miles, affording us an excellent opportunity of seeing the Railroad capabilities of this valley, which are excellent; presenting ground for a good line, with gentle grades, requiring a very moderate amount of work. Near the mouth of the Cowlitz we were met by another steamer belonging to the Oregon Steam Navigation Company, which carried us up the Columbia and Willamette rivers to Portland.

We were already in possession of the reports of Mr. Johnson, Chief Engineer of the Railroad Company, of the surveys made in 1867, of the different passes of the Cascade range of mountains, which intervenes between Puget Sound and the main valley of the Columbia, eastward, and therefore preferred examining on the ground the much more circuitous, though more level, route, via the Cowlitz river Portage, which goes around the mountain range, which is entirely cut through by the Columbia river. It will be observed that this is quite a distinct thing from what is called the Cowlitz Pass, which is by another prong of the Cowlitz river coming from the mountain. The surveys prove the practicability of routes through the passes of the Cascade range, as described in Mr. Johnson's report; and before the final adoption of a route, additional surveys will of course be made, covering carefully all the possible routes between Puget sound and Columbia river. It is unnecessary in this place to present more detailed views bearing upon the question of these routes. I can state that it is practicable to construct a good railroad between the Sound and the river at a reasonable cost.

On our return to Portland, the gentlemen of that city very kindly gave us all the information they possessed respecting the character of the country along the Columbia and Willamette, and the business of that region, and Captain Ainsworth, President of the O. S. N. Co., again placed at our disposal, free of expense, one of their fine steamers, which plies regularly between Portland and the Cascades on the Columbia river, a distance of sixty miles, which enabled us to have a perfect view of the great valley of the Columbia that far. On our way we touched at Fort Vancouver, a United States post and town of some importance, about five miles above the upper mouth of the Willamette river. At the foot of the Cascades we disembarked and were conveyed in the railroad cars, belonging to the same company, six miles to the head of the falls, where we were placed on board of another excellent steamer belonging to the company. The Cascades present to the tourist a charming scene, though to the engineer they are an obstacle; which has, however, been admirably overcome by the construction of a first-class railroad, which is traversed safely at the rate of twenty-five miles an hour; so that but a few minutes elapse between bidding adieu to one steamer and being comfortably quartered on another ready to proceed up the river, through the most picturesque scenery, fifty miles to the foot of the Dalles, at the City of Dalles, where another railroad, fourteen miles long, conveyed us to the head of the Dalles at Celilo, a small town, where we were accommodated with another of the company's steamers, which took us one hundred and twenty miles to Wallula, a small town at the mouth of the Walla Walla river. On our way we stopped at Umatilla, a town eighty-five miles above Celilo, and discharged thirty-three tons of freight in less than an hour.

At Wallula, we left the river, and proceeded into the interior thirty miles to Walla Walla, in Washington Territory, a town of considerable importance, within a mile of which stands new Fort Walla

Walla. There are many objects of interest to the tourist, such as the Cascades, the Dalles, with its great salmon fisheries, the Bridal-veil Falls, Castle Rock, Governor Stevens' Monument, and others which I have not here taken time to describe, although they are prominently known along the Columbia valley, and on Puget Sound, Mount Baker, Mount Adams, and Mount St. Helens, with their elevated snow-clad peaks, are the pride of Washington Territory; whilst Mount Jefferson and Mount Hood, with tips in the eternal snows, constitute grand and conspicuous objects in Oregon, and are visible, especially Mount Hood, at a great distance, from various points along the river. Were I to enter upon descriptions of all the wonders which attracted our attention, volumes would take the place of a report.

Thus, through the aid of the admirable steamers of the O. S. N. Co. we were enabled, in a brief time, to make a most satisfactory reconnoissance of the great valley of the Columbia, to within twelve miles of the junction of the Snake river with the Columbia; which is about three hundred and fifty miles from the ocean, and two hundred from tide-water in the Columbia river. On the greater portion of the way a good Railroad with low grades, can be built at moderate cost. There are some miles of heavy work, but my detail notes show that the miles of easy construction predominate so materially as to reduce the average cost within very moderate limits. One fact is of more value than many theories. The fact that a private company, in the very infancy of the white settlement of this portion of Oregon, constructed first-class Railroads around two of the most difficult points in the valley; one six and the other fourteen miles long, and thereby securing the control of traffic and passenger travel which has paid handsome dividends on the cost, is a practical proof of the feasibility of the route along the river.

Although our steamer stopped at Wallula, there is good steamboat navigation most of the year for more than a hundred miles farther up the Lewis or Snake river, and there are many long stretches of good steamboat navigation on the upper Columbia, and along Clarke's Fork.

The citizens of Walla Walla, and of the flourishing agricultural valley of which it is the business centre, have projected a Railroad to the Columbia river at Wallula; and, as the ground is quite favorable for the construction of a road of light grades, at very moderate cost, it will probably be built very soon. The enterprising farmers of that region will not long be contented with the onerous tax of thirty miles of common road to reach the grand commercial avenue—the Columbia river.

At Walla Walla we fitted out our horseback expedition for the journey across the country to the Rocky mountains, consisting of eight pack mules and ten saddle horses. During our short stay we were entertained by the hospitable people, and greatly interested in noting the abundant evidences of the wonderfully productive power of the valley, which is remarkable. Half a dozen years ago there was nothing here but the bare ground, where there are now elegant farms and gardens, in which almost every description of grains, fruits and flowers are growing in the greatest perfection and profusion. Philip Ritz, Esq., the new Marshal of Washington Territory, who kindly accompanied us during a large portion of our journey, has here a gem in the shape of house and grounds, orchards and gardens, all of which has been formed in less than five years. Mr. A. B. Roberts, an enterprising farmer, took us over his place and showed us agricultural productions, which, in quality, size and quantity, appeared most marvellous. The flower garden of Mr. Dugan, a prominent lawyer, was also perfectly charming, containing a great variety of the most beautiful flowers, which, in this far off land, constitute a pleasing contrast to the wilderness of grass, which, until recently, covered all this region and reigned supreme.

The latitude of Walla is about 46° 03′, longitude 118° 12′. The route from Walla Walla was chosen with a view to command, in the shortest time, the best general view of the country. We had before us the reports of Lewis and Clark, Governor Stevens, Captain Mullan, and others, and were accompanied by Mr. Ritz, a gentleman who has spent many years on the Pacific slope, and who has frequently passed through the mountains between the Columbia and the Missouri valleys. We therefore kept away from the immediate course of the river, and journeyed across the country on an intermediate route, leaving the Lewis or Snake river to the south, and Clarke's river to the north of us. We traveled at an average rate of about twenty-four miles a day, writing down as we advanced all objects of interest; our course being nearly northeast as far as Pend d'Oreille lake, one hundred and eighty miles from Walla Walla. We crossed a number of streams, the principal of which are the Touchet river, Lewis or Snake river, Palouse river, Camass creek, and Spokane river, which is the prolongation of the Cœur d'Alene river after passing through the lake of that name. The general character of all this region may be described as high rolling prairie, everywhere covered abundantly with bunch grass to the summits of the highest hills; treeless, excepting along the margins of the streams. Such is the country all the way to the northern boundary of the United States (latitude 49°), and beyond into the British Possessions, the Columbia river reaching by two of its principal tributaries as high as latitude 53°. The country near our northern boundary was explored by Captain McClellan in 1853. The information derived from these surveys and the surveys of Governor Stevens, and those of Captain Mullan and others farther south, together with our information derived from intelligent gentlemen who have been through this interesting region, added to our personal observations, leave no possible room for doubt; this is an immense grazing area, of the most superior character, interspersed with the valleys of perennial streams, along which are lands which when settled by industrious farmers will be of the most productive character, as we have seen in the case of a number of improvements already made; whilst the climate to the majority of mankind is not only salubrious, but remarkably attractive. In Washington Territory alone, along its eastern side, there are at least twenty thousand square miles, or twelve million eight hundred thousand acres of the finest grazing lands, on which thousands of cattle and sheep will be raised as cheaply as in any other quarter of the globe; and this grass is so nutritious that the cattle raised upon it cannot be surpassed in their weight and quality. Snow rarely falls to a sufficient depth to interfere seriously with the grazing all through the winter. Such may be taken as the general view upon this important point respecting an area of territory nearly half as large as the State of Pennsylvania.

The Lewis or Snake river valley and the valley of the Spokane, have been referred to in reports of the Company as routes for the Railroad, and before a-final location is made they will, doubtless, receive further examination.

Our route was taken across the valleys, and we passed from Washington into Idaho Territory. At Lake Pend d' Oreille, (in Idaho), our animals, etc., were put on board the steamer of the O. S. N. Co. which had been placed at our service by Captain Ainsworth, and were conveyed in three hours along the eastern coast to the inlet of Clarke's river, a distance of thirty miles, where we landed our animals at an Indian village on the right bank at Patrick's Landing. We then proceeded with the steamer to explore the Lake, continuing around its eastern and northern side, and returning by its southern and western coast, in all, about sixty miles, concerning which I have numerous notes in detail. It is about thirty miles from the entrance of Clarke's river to its outlet. Lake Pend d'Orielle is a lovely body of water, perfectly clear and fresh, and profound in its depths, which are yet unfathomed. It is shaped irregularly, in the form of an ear-drop, as the name betokens; each pendant being about thirty miles long, and from three to seven miles wide; covering a total area of about three hundred square miles, or one hundred and ninety-two thousand acres. It abounds in fish, and presents some of the most charming natural scenery, peculiar to itself. A great portion of its eastern shore consists of high, abrupt rocks, broken down here and there by the valleys of small streams entering into the lake. The country on its western and northern sides is more flat, and better adapted to Railroad construction. The most northerly of the routes proposed for the Northern Pacific Railroad, passes down the right bank of Clarke's river to its point of debouching into this Lake, and around its north shore to the outlet of the same river near its northwestern end. Thus Clarke's river, which is a very large stream, draining an immense area along the western slope of the Rocky Mountains, runs directly through Lake Pend d'Orielle, maintaining it in perrennial salubriousness. Should the Railroad be built by this route, this lake will become as famous among tourists as the romantic lakes of Switzerland.

Leaving the lake inlet of Clarke's river, our march was pursued up that stream along its right bank, passing, about twenty miles up, the "Cabinet Rocks," to which point the O. S. N. Co's lake-steamer runs during the spring and early summer, carrying packers and their animals with merchandise for Idaho, Utah and Montana Territories, coming from Portland, Oregon, and other places along the Columbia river valley. The present season, 1869, has been very unfavorable, owing to an unusual drought, which has prevailed all over the mining regions and along the entire Pacific slope in the State of Oregon, and in Washington Territory. About four thousand animals and their packs were conveyed this season, but many packers declined availing themselves of the facilities of the steam navigation, taking their trains around the northern end of the Lake by a circuitous trail difficult to pass in wet seasons. We observed many striking and interesting objects on our route along the valley of Clarke's river, but I am warned by the present length of these preliminary observations, and cannot here introduce a description of them. The Cabinet rapids, the zig-zag trails along steep bluffs, fifteen hundred feet above the river, distant views of high peaks, grand forests, splendid prairies, etc., are parts of these. Our course was continued along the right bank of the river as far as the junction of the *Flathead* and *Bitterroot* rivers forming Clarke's river. Both of these are large streams, always containing a great deal of water. Railroad lines have been projected along them, and their valleys have been explored. They are known to present practicable Railroad routes. Soon after leaving the Lake we entered Montana Territory.

It is about one hundred and thirty miles from Lake Pend d'Oreille, along Clarke's river to the Flathead river, upon which distance there is a variety of scenery; some of the grandest and some of the most beautiful to be found anywhere. For a railroad valley, as a whole, it is not merely favorable but attractive. The few points of difficult work along the rocky bluffs, when thrown into the aggregate, will not increase the cost so as to raise it above a moderate sum per mile; while the grades will be very light, and the curvature generally easy. Clarke's river has a flow in low water at least six times greater than the low water flow of the Ohio river, between Pittsburgh and Wheeling, and while its fall is slight, considered with reference to railroad grades, it is so considerable as to afford a great number of water-powers, whose future value must be very great; an average of eleven feet per mile.

Around Lake Pend d'Orielle, and for some miles westward, and all along Clarke's river above the lake, as far as we traversed it, there is a magnificent region of pine, cypress, hemlock, tamarack and cedar timber, many of the trees of prodigious size. I measured one which was thirty-four feet in circumference, and a number that were over twenty-seven feet, and saw hundreds as we passed along that were from twenty to twenty-five feet in circumference, and from two hundred to two hundred and fifty feet high. A number of valleys containing large bodies of this character of timber enter Clarke's river from both sides, and the soil of these valleys is very rich. Clarke's river valley itself is for much of the distance confined by very high hills approaching near to the stream in many places; but there are sufficient sites for cities and farms adjacent to water-powers of the first-class; and not many years can elapse after the opening of a railroad through this valley till it will exhibit a combination of industries and population, analagous to those which now mark the Lehigh, the Schuylkill, the Susquehanna, and the Pomroy region of the Ohio river. Passing along its quiet scenes of to-day, we can see in the near future, the vast change which the enterprise of man will bring. That which was once the work of half a century is now the product of three or four years. Indeed, in a single year after the route of this Northern Pacific Railroad shall have been determined and the work fairly begun, all this region, now so calm and undisturbed, will be teeming with life instilled into it by hardy pioneers from the Atlantic and from the Pacific.

Passing along the Flathead river for a short distance, we entered the valley of the Jocko river. The same general remarks concerning Clarke's river valley are applicable to the Flathead and Bitterroot valleys. The climate, the valleys, the timber, the soil, the water-powers; all are here, awaiting only the presence of the industrious whiteman to render to mankind the benefits implanted in them by a beneficent Creator.

We passed up the Jocko valley about thirty miles, and thence by a favorable divide over to the waters of the Hellgate river, a very fine stream, the main continuation of the Bitterroot. Our measurement of the flow of the Hellgate (August 10, 1869,) gave 115,000 cubic feet per minute, when the stream was considered to be quite low. The distance by our travelled route from the Flathead ferry to the town of Missoula, on the Hellgate river, is about fifty-six miles. On all of this it is easy to build a railroad at moderate cost; timber and stone being quite convenient; although for a portion of the way it is a nearly level, treeless prairie.

The people of Missoula gave us a warm welcome, and we have to thank them for many kind attentions. Before arriving there we were joined at our camp on the Jocko by Major John Owen, of Fort Owen, who rode sixty miles to meet us, in company with Major Graham, of Deer Lodge City. These gentlemen gave us much useful information concerning the region through which we were passing, and accompanied us to Missoula and Deer Lodge City.

Missoula is a thriving young town near the western base of the Rocky Mountains, containing a grist-mill, saw-mill, two excellent stores, and twenty-five to thirty dwellings, a number of them well built. I visited McWhirk's garden of five acres, where I found ripe tomatoes, water-melons, musk-melons, remarkably fine potatoes, beans, onions, peas, and squashes; also young apple trees and other fruit trees, and a very fine collection of flowers; and all this had been brought about from the virgin soil in two years, and would this year yield the owner over two thousand dollars gold, the only currency yet known in that part of Montana Territory.

Messrs. McCormick, Warden, Higgins, and other gentlemen of Missoula, did all in their power to aid us in procuring satisfactory information of various kinds.

It was gratifying to find manifested here, and at every point where there were settlements, the most intense interest in the speedy opening of the Northern Pacific Railroad; which they regard, and very properly, as the only means of placing Montana Territory in a proper position among the States of the Union, and of permanently establishing its general prosperity.

We were accompanied from Missoula by Messrs. Worden and Higgins, who remained with us a number of days, aiding us materially in our examinations.

Just after leaving Missoula, by the way of Mullan's military road, we entered what is called "the gate of the mountains," a romantic, picturesque pass, where the Hellgate river cuts through a mountain. We crossed the Blackfoot river five or six miles from the town, on a bridge with four spans of seventy-five feet each. We made the flow, by measurement, then in the Blackfoot, about 48,000 cubic feet per minute—at its lowest stage.

My notes of our route over to Deer Lodge City, eighty-five miles by our route from Missoula, are very full; but I can only take space here to state that our examinations satisfied me that the routes by the way of the Blackfoot, and the Little Blackfoot, and the valley of the Hellgate, to points near their respective passes on the summits of the Rocky Mountains, are quite practicable. These passes will be referred to further on more particularly.

Deer Lodge City is a flourishing young city, situated in a remarkably fine valley, through which flows Deer Lodge Creek, the principal fork of the Hellgate. When we were yet ten miles off from this place, we were met by a large delegation of the prominent citizens in a number of carriages and on horseback, who were quite enthusiastic in view of the promise of the Railroad, which seemed to them to be foreshadowed by our coming. Whilst in this interesting locality, which is one of the stirring business centres of the gold region, we were the recipients of many attentions, and obtained a great deal of valuable information. I may with propriety refer more particularly to Granville Stuart, Esq., who has made this whole region a study for years, and to Major Blake, Major Graham, and Judge Dance, who accompanied us in our examinations of the Deer Lodge Pass and other places.

The Deer Lodge valley is very wide in places, ten to fifteen miles from the hills on one side to the hills on the other; nearly level, laterally, and everywhere clothed with rich grass, upon which we observed numerous herds of tame cattle and horses feeding. The Deer Lodge creek flows through it in a meandering direction and adds immensely to its value as an agricultural region. Some farms are cultivated, but farming is yet in its infancy, and there are thousands of acres of arable land here and elsewhere in Montana awaiting settlement.

A few miles above Deer Lodge City there is a very remarkable mound, called the "Deer Lodge," from which the valley, the stream, and the city derive their name. It stands a very prominent object in the midst of a wide, flat valley, and has been formed entirely by the accretion from the unceasing flow of hot springs, two of which exist now on the very summit, which rises thirty-three feet above the plain. It is thirty feet in diameter at the top, and a hundred and thirty at the bottom, surrounded by numbers of springs of various degrees of temperature, the hottest of which we found to be 160°, and the coolest of eight that we tried 98°. The material resembles a light iron ore of a very curious formation. It is a very great curiosity.

Considered as a railroad route this valley is remarkably favorable, the rise from Deer Lodge City to the pass or divide between the waters of the Pacific and Atlantic, being quite gentle;

and even on the last few miles, the summit, about five thousand feet above the sea, may be attained without employing a gradient exceeding fifty feet to the mile, with a moderate cut. The whole forty miles from Deer Lodge City to the summit of the Rocky Mountains by this route can be built as cheaply as roads are built through prairie countries generally. A little more work will be required in passing on the east side from this side down Divide creek to Wisdom or Big Hole river; but the line will be highly favorable as an average all the way to the Jefferson Fork of the Missouri river. This favorable Pass comes into connection more particularly with the Yellowstone valley route to the main Missouri valley. A remarkable circumstance connected with this Pass will convey a very clear view of its peculiarly favorable character. Private parties engaged in gold mining, in the gold field which exists abundantly on both sides of the Rocky Mountains, have dug a ditch across this summit which is only eighteen feet deep at the apex of the divide, through which they carry the water of "Divide creek," a tributary of the Missouri, across to the Pacific side, where it is used in gold washing, and the waste water passes into the Pacific ocean. This has been justly termed highway robbery. The route running down the Jefferson Fork, crossing the Madison Fork over to the Gallatin, and up that valley to near the Bozeman Pass, is very favorable, admitting of easy grades and curves at moderate cost. The Bozeman Pass is about five hundred feet lower than the Deer Lodge Pass, or about four thousand five hundred feet above the sea. Some heavy work of grading occurs on both sides.

The Bozeman divide is not so favorable as that at the Deer Lodge summit, from the fact that the ascent to it on either side is less gentle, though, in comparison with other passes of the mountains, it is quite favorable, being practicable without the aid of a tunnel, with no more costly approaches.

The Yellowstone valley, which we saw only for a short distance, in connection with our examination of the Bozeman divide, is known to be advantageous as a route for a great trunk railroad between the East and West. The surveys of General Raynolds, made in 1860, established this fact, and quite recently the expedition under General Hancock has confirmed it. General Hancock reports that the Yellowstone valley is favorable for the construction of a Railroad, and that there are large bodies of valuable land there, needing only the Railroad to develop them. From the mouth of the Yellowstone, eastward to Lake Superior, the capabilities of the country are familiar to many, and known to be, as a whole, very favorable for the cheap construction of a Railroad. I am, therefore, inclined to the opinion that the most advantageous line, in certain important respects, will be found to be by the way of the Yellowstone valley, Bozeman's Pass, Gallatin river, to a point convenient for crossing over to the Madison, and across that stream to the Jefferson, up the Jefferson to the Deer Lodge Pass, and down the Hellgate river. Thence westward there is a choice of routes; one passing down Clarke's river and around by Lake Pend d'Oreille, and thence across the great plain to the Columbia, at or near the mouth of Lewis or Snake river; the other an alternate route, down Hellgate to the Bitterroot, along the Bitterroot to its Lou Lou Fork, up that fork to the divide in the Cœur d'Alene range of mountains, and down by the most convenient fork of the Clearwater river to Lewistown, at the junction of the Clearwater with Lewis or Snake river, and thence down Snake river to its mouth, there meeting the line first above described—the Pend d'Oreille line.

This general route (by either of these two lines) may or may not prove to be the shortest on a final location; but it will as a whole be on the route of the least snow; the Deer Lodge Pass of the Rocky Mountains being noted for the comparatively small quantity of snow which falls or lies there. Captain W. W. DeLacy, who has spent many years in surveys in this region for the government, and who was connected with the surveys and reports of Captain Mullan, as well as those of Governor Stevens, stated to me, as his opinion, founded on personal experience and an excellent knowledge of the topography and climate of Montana that the line of easiest grades and least snow passing through

Montana Territory, would be by Bozeman's Pass from the Yellowstone valley, and through the Deer Lodge Pass, substantially as I have described it. The line of "easiest grades and least snow," crossing the Rocky Mountains, can afford to encounter some increase of distance, especially when to that can be added, *least cost per mile for construction.*

Returning from our examination of the Deer Lodge Summit we passed over to the Little Blackfoot river, down which the line must run in case Mullan's Pass should be chosen for the route over the Rocky Mountains. It may as well be stated here that the Little Blackfoot affords a good route for the road westward after the mountain has been passed, and that the Blackfoot offers a good route for a line from Cadotte's Pass, after the mountain has been overcome from the east. My examination of both of these streams at their sources and at points some miles below, in conjunction with reliable information from Captain De Lacy and others, enable me to offer the above statement; but the difficult portions of these routes are caused by the topography east of the mountain. Tunnels are necessary at both of these passes.

Mullan's Pass, which is about sixty miles northward of the Deer Lodge Pass, is a little over six thousand feet above the sea, or one thousand feet higher than the Deer Lodge Divide. It is approached from the east by the Valley of the Missouri river by one of the branches of the Prickly Pear, a tributary of the Missouri. According to the survey and report of Captain Mullan it would require a tunnel two and a fifth miles long; and from an examination of the place it appears to be naturally adapted to tunneling—the ground falling off abruptly, especially at the eastern approach. Excepting that a tunnel of that length is required, the line of the general route up the Missouri Valley and down the Little Blackfoot on the west side is fairly practicable at comparatively reasonable cost when the mountain portion cost is distributed among the aggregate miles.

After examining the crossing of the mountain at Mullan's Pass, we proceeded to Helena, now the principal business centre of the Territory of Montana, where we were received with great cordiality and many demonstrations of interest in the fortunes of the Northern Pacific Railroad. The new Governor, Hon. J. M. Ashley, happened to arrive at the same time, and made his reception address to the people the same evening; and in that he took occasion to lay great stress upon the paramount importance of the Northern Pacific Railroad to the people of Montana. We were indebted to many gentlemen for attentions and information received in Helena; S. T. Hauser, Esq., N. Langford, Esq. and many others, gave us much information and much kind attention.

From Helena we went in a coach over a good road along the Missouri river valley, one hundred and fifty miles, taking notes of the country going and returning. We stopped at Fort Shaw, on Sun river, and after a pleasant interview with General De Trobriand, obtained from him the promise of an escort, for the purpose of visiting Cadotte's Pass, at the source of the middle fork of the Dearborn river. We spent a day at Fort Benton, examining the points of interest in that vicinity, which were shown to us by the gentlemen of that place. Among other things we looked at the out crop of a coal mine which had recently been discovered within less than two miles of the place. This has been an unfortunate season for the business men of Fort Benton; the water has been so low in the Upper Missouri river that the steamers could not ascend that high, and were compelled to discharge their cargoes several hundreds of miles below.

Returning to the crossing of the Dearborn river, we found our escort arrived in advance of us. The next morning we started on horseback, up the Dearborn river, guided by Mr. Caldwell, whom we brought from Fort Benton, and Mr. Cooper, who is working a coal mine at this point, on the Dearborn river. We passed across several forks of the Dearborn, and were on some very elevated points, affording us a good view of the topography which was spread, map-like, before us, unencumbered by trees, though covered with some of the richest grass we had seen anywhere during our long journey. In the afternoon we arrived at Cadotte's Pass. I went down about two miles on the west side to the waters of the Blackfoot, to where I could see the character of its valley, which below here is favorable for a Railroad. This Pass was surveyed under the direction of Governor Stevens, and its general characteristics are known. It is not materially different in its elevation from Mullan's Pass, and like that has abrupt ascents on both sides; but a much longer tunnel is unavoidable. The approach on the eastern side, from its appearance, is much more costly than that from the western side, on account of several heavy ravines, and because the country on the eastern side is 567 feet lower at the proposed tunnel entrance than the western. Captain Mullan, who surveyed this Pass, thus refers to it in his report:

"The tunnel will be 4.19 miles in length, when there will be a cut commencing fifty feet deep, and coming out in the valley of the Blackfoot five-hundred and fifty-eight thousandths of a mile west of the tunnel, [nearly six-tenths of a mile.] The cut and the tunnel have a grade of sixty feet to the mile. The highest point of the road, therefore, will be at the entrance of the cut, an elevation of five thousand one hundred and ninety-five feet above the level of the sea, and eight hundred and forty-nine below the mountain summit. [Summit of ground 6,044 feet.] The entrance of the cut will be two and nine hundred and fifty-eight thousandths miles [2.95 miles] west of the western base of the mountain, which is five hundred and sixty-seven feet higher than the eastern."

Captain Mullan's line approached the eastern end of the tunnel with 60 feet per mile grades. By the use of 80 feet per mile grades the tunnel on his line could be started 200 feet higher, and would be considerably shortened; but it is obvious that this portion of the route would be very costly and difficult.

Captain Mullan also surveyed a route through Lewis' and Clark's Pass. This, as well as the other, was traced from the Missouri, via Sun river, and on the last nineteen miles approaching the tunnel, he reports "very difficult and heavy work.

The tunnel itself is 2½ miles long. The summit level of the tunnel is at a debouche a half mile west of the western base of the mountain, and is 5,698 feet above the sea." "The mountain pass being 6,519 feet above the sea." The difficulties on the Lewis and Clarke Pass route, as surveyed, are so great as to render it almost impracticable in view of the fact that there are other routes more advantageous. We then proceeded from Helena to Bozeman City, where we were well received by the gentlemen of that place, and there met a delegation of gentlemen from Virginia City, who accompanied us over to the Yellowstone. Col. Brackett, who is in command of Fort Ellis, about three miles from Bozeman, kindly furnished us an escort under the command of Lieut. Hamilton, who went with our party over the Bozeman Summit and back.

It will be impossible for me in the space and time contemplated for this preliminary Report to consider thoroughly, and properly discuss in detail, the merits of the several routes proposed for the Road crossing the Rocky Mountains; enough, however, is known to enable me to offer an approximate estimate of the cost, etc., of constructing a Railroad, of about two thousand miles length, between Lake Superior and Puget Sound, by the route of the Missouri river valley part of the way, and by the Yellowstone valley from its mouth, via Bozeman's Pass, as already described.

The country westward from Lake Superior to the Missouri river has been so frequently explored that its characteristics are now well known. They are ably and clearly set forth in the Reports made to you, a few days ago, by Governor Marshall, of Minnesota, and Philip W. Holmes, Esq., which I have studied.

PROBABLE COST OF ROAD.

I present an approximate estimate of the cost of building the Railroad, at present ruling prices for work, from Lake Superior to the head of Puget Sound, a distance of two thousand miles, on the longer of the proposed routes, as nearly as can be ascertained from former surveys and explorations, and from my own personal observation on a large portion of the line, including all the most difficult parts.

It is proper to preface this estimate with some explanatory remarks. The line upon which the estimate is to be given runs from the head of Lake Superior across the Mississippi, the Red river, and the Dakota river to the Missouri; thence crossing the Missouri into the valley of the Yellowstone, and along that stream to Bozeman's Pass, through the Belt range of mountains; thence down the Gallatin valley, crossing the Madison river, and over to the Jefferson valley, and along that to the Deer Lodge Pass of the Rocky Mountains; and along Clarke's valley to Lake Pend d'Oreille; and from the lake across the Columbia plain to Lewis or Snake river; down that to its junction with the Columbia; along the Columbia river to the Cowlitz river; up the valley of the Cowlitz; and over the portage between the Puget Sound and Cowlitz river, and down to Puget Sound at its southern extremity, whence the road may be carried along either side or both sides of the Sound, as far as may be desired, to any port or ports which shall be selected.

Although I would not feel prepared without having the results of further surveys to pronounce this the best possible route which can be found between Lake Superior and Puget Sound, it certainly presents important advantages, and is known to me to be eminently practicable. Shorter routes may be traced, but probably none which will be cheaper, per mile, or which will offer so good a profile for profitable service as a great main trunk Railroad thoroughfare.

In making this estimate, I assume that the graded road bed, bridges, culverts, etc., are to be such as we find on our first-class roads; and that the track is to be thoroughly constructed, with rails of sixty pounds per lineal yard, put together with the most improved joint-ties, and completely ballasted with gravel or broken stone. On the prairie portions, which constitute more than half of the whole road, the track cannot be ballasted at the time of laying; experience has proved, however, that new roads over such level plains can be run with entire safety at the usual rates of speed, previous to receiving the final ballast. Sufficient allowance is made in the estimate for the cost of conveying the material for the track; though this item will be much less on this road than might at first appear, on account of the number of points where the line crosses, or runs parallel with, navigable water courses, along which the iron rails, cross-ties, etc., can be cheaply transported. The Northern Pacific Railroad route is in this respect very remarkable. Thus, with the magnificent lake navigation to begin with, there is the Mississippi river, the Red river of the north, the Dakota river, all navigable streams crossed at convenient intervals along the first five hundred miles of the route. Thence along the immediate valley of the navigable waters of the Yellowstone river, four hundred miles, to within twelve miles of Bozeman's Pass, in the Belt range. Thence to the navigable waters of the Flathead river, a distance of only about two hundred and twenty miles, which includes the passage of the Rocky Mountains, no part of the line will be more than ten to fifteen miles away from good timber; and beyond that, along the valley of Clarke's river, it will run through or adjacent to forests in which everything in the shape of wood required on railroads is to be had for the cost of cutting. From Clarke's river to the Pacific Ocean, there is cheap water communication all the way. It is clear, therefore, that these striking and advantageous characteristics of this line, conferred by the water courses, and which are equally applicable to a line by the Missouri river and through the more northerly mountain passes, render it unnecessary to add a large per centage for the extra cost of carrying track materials.

For convenience of reference I have arranged the total length of the line to be estimated into six divisions, which appear to be natural, and put the estimate in tabular form.

APPROXIMATE ESTIMATE.

No of Div.	DESCRIPTION OF DIVISION.	Length, Miles.	Estimated Cost.
1.	Lake Superior to Yellowstone river,	550	$13,750,000
2.	Along the Yellowstone to Bozeman's Pass, ...	420	11,760,000
3.	Bozeman's Pass to Hellgate river, Mountain Division, . . .	225	9,000,000
4.	Hellgate river to Pend d'Oreille lake, ...	205	7,000,000
5.	Pend d'Oreille lake to the mouth of Lewis river, ...	223	7,500,000
6.	Mouth of Lewis river to Puget Sound—Columbia valley Division, ...	377	11,310,000
			$60,320,000
	Add for sidings and additional track, . . .		4,200,000
	Contingencies, including superintendence and engineering, 10 per cent., ...		5,000,000
	Telegraph line and instruments, complete, $300 per mile, ...		600,000
			$70,120,000

	BUILDINGS.		ESTIMATED COST.
134	Wood and water stations,	@ $3,500	$469,000
20	Engine-houses and turn-tables,	@ 15,000	300,000
5	Principal engine repair-shops,	@ 100,000	500.000
2	Principal car repair-shops,	@ 75,000	150,000
5	Principal car repair-shops,	@ 40,000	200,000
200	Section, tool, and hand-car houses,	@ 500	100.000
134	Freight and passenger stations,	@ 2,000	268,000
150	Freight platform stations,	@ 500	75,000
10	Principal freight and passenger depots,	@ 25,000	250,000
			$2,312,000

	ROLLING STOCK, ETC.		
120	Locomotive engines, freight and passenger,	@ $13,000	$1,560,000
100	Passenger cars, first class,	@ 4,000	400,000
50	Passenger cars, second class,	@ 2,500	125.000
30	Smoker's cars,	@ 3,000	90,000
30	Baggage, mail, and express cars,	@ 2,000	60,000
1500	Box, freight, cattle and platform cars,	@ 800	1,200,000
40	Caboose and wrecking cars,	@ 1,200	48.000
20	Tool cars,	@ 800	16,000
80	Hand cars,	@ 200	16,000
	Tools, snow-ploughs, etc. (per 100 miles),	@ 1,000	100.000
			$3,615,000

In addition to the foregoing there are various outlays, which will necessarily attach to the work during its progress, not covered in the items given, or included under the usual per centage allowed for contingencies; especially at the principal terminal points on Lake Superior and Puget Sound, and on the Columbia river, and likewise at the crossings of the Mississippi, the Red river, the Dakota river and the Missouri river, to connect the Railroad business conveniently with the transportation to and from these respective streams.

Also on the route on which this estimate is predicated, there would be a branch line, a few miles in length, extending to Portland, Oregon. In case the line should be along the Columbia river at Fort Vancouver, it would be only about five miles across to Portland, but involving two costly bridges, one over the Columbia, the other over the Willamette river.

In order to cover the cost of such a branch and the cost of the necessary extra works above mentioned, it will be proper to add to the general estimate the sum of $1,200,000 for the branch, and $800,000 for the extra works referred to; making two millions in all. Nothing has been inserted for "right of way," as the land-grant carries with it all that is needed over nearly every foot of the way; and where land is taken up, the owners will gladly give the Company all they may require, in consideration of the benefit to the remainder.

RECAPITULATION.

Grading, masonry, bridging, track and ballast,	$60,320,000
Sidings, etc.,	4,200,000
Contingencies, including superintendence and engineering,	5,000,000
Telegraph line,	600,000
Buildings,	2,312,000
Rolling stock,	3,615,000
Branch road,	1,200,000
Extra works, etc.,	800,000
	$78,047,000
Interest on bonds over receipts during construction,	7,230,000
Total,	$85,277,000

This gives an average of $42,638 per mile.

It is obvious that if this estimate is sufficiently liberal to cover all probable contingencies, that the road can be opened for the beginning of traffic for a less sum than the above total; for it is scarcely presumable that all the works at the terminal points, or the ballasting, would be completed at that time, and it would not be necessary to have all of the buildings or rolling stock until business on the road called for them.

I have purposely made no allowance for the reduced cost which may be brought about by the introduction of Chinese labor. It is well known that a large amount was saved in the construction of the Central Pacific and Union Pacific Railroads by the employment of Chinese. It would be more or less hypothetical to assume on this account an important reduction of the amount of the estimate, and I prefer that yourselves and others should judge of it.

My estimate is materially less than estimates which have previously been submitted to the public. Mr. Johnson, the able Chief Engineer of the Company, who has made this route a study for nearly twenty years, and whose original report, published in 1854 (fifteen years ago), I regard as one of the most masterly engineering productions of the present age, obviously designed that the estimates, based, as they had to be, largely upon hypothetical views, should be ample.

I have personally examined and carefully noted all the more difficult portions of the route, excepting such as there may be along the Yellowstone valley; and in regard to that, the report of General Raynolds, published in 1868, and the very recent report given to yourselves by Major-General Hancock, who made a reconnoissance of the Yellowstone valley during the present year, go to confirm the opinion that my estimate is safe for the four hundred and twenty miles assumed between the mouth of the Yellowstone and Bozeman's Pass; namely, an average of twenty-eight thousand dollars per mile, exclusive of the buildings and rolling stock.

It is not possible to embody in a brief special report the numerous notes of the country made by me during the past summer, which are the basis of the statements herein offered; nor can the advantages of the general route of this road and its future prospects be more than glanced at without more elaborate remarks. It would be unjust to all parties not to refer more particularly to certain cardinal features which the general route of the Northern Pacific Railroad offers for consideration, both in its relation to the internal trade of an immense area of our country, and as the most convenient line for accommodating the commerce of Asia, and of the islands of the Pacific ocean.

In my opinion an increasing commerce with Asia and with foreign countries in general, with the city or cities at the western end of this Railroad, will have the effect of very rapidly augmenting the population on the Pacific slope, not merely or principally by immigration from Asia, but chiefly by emigration across the continent—the overflow of the redundant population of the Atlantic States and of Europe. The peopling of these vast areas in the Columbia valley, abounding in the elements which will yield a liberal support to millions of inhabitants, will open up an entirely new field for the world's industry, thus adding largely to its general trade and commerce. The time seems to have arrived when, in the order of events long ago designed by an overruling Providence, this glorious valley should be handed over to the ameliorating influences of civilized population. The buffalo has already disappeared, and is no longer found throughout the greater portion of the territory to be commercially accommodated by this line of Railroad; the Indian, whose nature and habits make him dependent largely upon hunting, is gradually receding as the white man's path widens, and his power establishes that dominion which it is vain for the Aborigines to oppose. It is no longer merely the penetration, by toilsome marches, of a few half protected settlers along a slow moving line of progress, the two oceans are now bound together by the line of Railroad already completed, the forerunner of other continental east and west Railroads, which are the only means by which the country can be developed, and by which any intrinsic value can be given to the government lands; each line of Railroad will take in its own special area, from which the Indian must recede, giving place to the very different industrious elements which belong to a higher order of our race.

The Northern Pacific Railroad route is advantageously situated for the early development of a very extensive area, reaching far into the British Possessions on the north, and presenting a clear field to the south of millions of acres of land adjacent to it, to be made a feeder to this line by means of branch roads.

The valley of the Red river, which runs almost due north into Canada, embracing one of the finest wheat regions in the world, will of itself forever insure to the eastern end of the road a profitable trade; and the construction of a north and south railroad through the Red river valley, connecting the main trunk with the region around Lake Winnepeg, will add largely to the business of the Northern Pacific line. The elaboration of isothermal lines has shown that the northern boundary of the United States, latitude 49°, an imaginary line, instead of being the northern boundary of cultivable lands and habitable climate, runs south of a vast body of very superior arable territory, only needing railroad facilities for its successful development. Extensive settlements are already there, knocking at the door, asking American enterprise to open it and unite their commercial destinies with ours. The age of railroads has sealed the doom of political lines of demarkation, and the progress of events is gradually but surely disseminating on this continent that spirit of self-government, the sure offspring of increasing popular intelligence, which must eventually, and perhaps very soon, remove the frail barriers which now separate the United States and Canada as governments; but whether the two countries shall or shall not speedily unite their political government, it is certain that the people will unite in efforts to develop the region lying along the present boundary on both sides between the two countries, and thus practically remove it. On the other hand, southward of the line of the Northern Pacific Railroad route, it is from three to five hundred miles to the line of the present completed railroad between Omaha and Sacramento. Granting that half of this area may find its most convenient outlet by that road, there are left of this space about two hundred and fifty thousand square miles, or one hundred and sixty millions of acres, in Minnesota, Dakota, Montana, Wyoming, Idaho, Oregon and Washington; destined at no distant period to contain a population of many millions.

These vast interior regions, without the essential aid of railroads, would remain for very many years in the future, as they have through long ages in the past, entirely valuelesss as marketable property of our government, but with railroads in operation, affording first-class facilities for the enterprising from all countries, the entire face of the land will quickly be changed from the virgin fields and woods to the civilized improvements of mankind, which alone can confer practical values in the world. Thus, by the completion of railroads, and in no other way, can the Government of our people, which is the property of every citizen, realize any pecuniary benefit from these interior Territories and States. Our people, therefore, cannot err in granting lands, privileges and credit to enterprising capitalists who are willing to employ their money, their time and their talents in the execution of these grand avenues of commerce; for the construction of such a line as this between Lake Superior and the Pacific, will give actual value to a region, on an average, at least thirteen hundred miles long by four hundred miles wide, or three hundred and thirty-two million eight hundred thousand acres, which, without the transportation facilities which this line will afford, would, for the most part, rest undisturbed in its primeval condition, comparatively useless and valueless to the government and to our people.

When the granting of a few millions of acres to a Railroad company, carries with it the immediate enhancement, or rather a real creation of value upon six times that quantity of government lands, the soundness of the policy which dictates the grant is surely beyond question.

There is another point which now assumes importance. In the earlier days of discussion concerning Railroads across the continent, it was a matter of moment how to bring about their construction without extravagant cost, in any reasonable number of years; but this, like most great questions which arise in the movements of mankind, whilst unceasingly advancing in myriad forms of developement, finds a solution in the accompanying circumstances. Thus in the construction of the Union and Central Pacific Railroads, the *need* led to the *way* of building miles of railroad in a marvellously short time.

This experience will of course be applied to the Northern Pacific Railroad; and with the remarkable natural facilities already mentioned arising from its junction with so many navigable streams, and with Lake Superior at one end and the Pacific Ocean at the other, there is no reason why, instead of occupying half a dozen years it should not be completed in *three*. I can now clearly see how it can certainly be built and opened for use between Lake Superior and Puget Sound in 1873; even allowing the greater part of a year to be devoted to surveys before determining in all cases the precise route; but meanwhile having all the preliminary plans and preparations thoroughly digested and arranged, and some portions begun.

THE LANDS.

In the grant of land of alternate sections on a width of forty miles on each side through the Territories, and twenty miles on each side through the States of Minnesota and Oregon, it is assumed that there may be nearly fifty millions of acres which will become the property of the Company. You are already familiar with the general excellent character of the land in Minnesota and Wisconsin, and from the reports of Governor Stevens, and more recently the report of Governor Marshall and Philip W. Holmes, Esq., who under your auspices made a special examination of the country as far as the Missouri river during the present season, you cannot but feel satisfied that a Railroad between Lake Superior and the Missouri river will, within a brief period after its completion, receive local traffic and travel enough to maintain it. Now this can only arise in consequence of the settlement and cultivation of the lands along the route; and it is therefore obvious that by a judicious system of land sales, settlers can be induced to occupy this territory at once, taking on favorable terms portions of the Company's lands, thus adding largely to the value of the residue, which can be sold on a gradually rising scale of prices, naturally induced by increase of population and the resulting industries.

I must not either designedly or undesignedly convey an impression that all of this immense body of land is arable, or that it is all adapted to the raising of wheat. There is in the large area embraced in this extensive land-grant a great variety of soil and climate, and through the mountain region there is a considerable portion, which but for its mineral wealth would have little value. There are tracts where, owing to the general absence of sufficient rains and dews, the land cannot be made to yield well without irrigation. On the other hand there are numerous valleys of rich land which will yield abundantly, at once, without artificial irrigation; and there are many millions of acres of the finest timber in the world. It is of course a question of time when this region shall be settled and sending forth supporting Railroad business; that time will be regulated wholly by the period when the Railroad shall be put in operation. If the commencement and completion of the road were to be deferred for ten years, the whole country would be very nearly ten years later before it could stand, as it should, with the rest of our American civilized regions.

At present the middle portion along and on both sides of the Rocky Mountains, is wholly dependent on the gold production, which has been very remarkable. Gold exists over a very large area in Montana Territory, also in Idaho, and parts of Utah and Oregon and Washington Territories. Gold has been the prime cause of the movement of the people to that once distant interior, but which the opening of the Railroad to the Pacific has brought comparatively near. Only a few years have elapsed since Montana became the theatre of the gold excitement, yet already a number of farms are established, and the capabilities of the soil and climate are not now a matter of conjecture, but of actual proof. Wheat, butter, nearly all kinds of vegetables, which, until recently, brought fabulous prices all through the gold region, are now but little higher proportionally than in the old settled parts of the States; and this has been effected wholly by the cultivation of farms in Montana Territory. I have the notes, and could give much more detailed views on this and other important points; but were I to enter upon the subject in that way, it would change entirely the character of this special report, rendering it too voluminous. Yet I desire to convey to others the impression that I have myself received from personal observation respecting the capabilities and the drawbacks appertaining to these lands. The entire range from the Missouri river to the Columbia river at Wallula is a grazing region, for the most part covered with very nutritious bunch-grass, upon which wild animals have fattened for ages, and upon which hereafter immense herds of tame cattle will profitably be raised. As a rule, all over this wide expanse of grass-clothed plains and hills there is an inadequate supply of rain, although the droughts are never sufficient to destroy the bunch-grass pasture; and, as a rule, the soil is excellent, and in many places very rich, needing only irrigation. There are, however, numerous valleys with perennial streams fed from the Rocky Mountains, containing land of the most fertile character, already improved in various places, and yielding abundant harvests. Lands, which if left without Railroad facilities, might be allowed to remain forever as natural pasture, will, when there are chances for shipment of the product by rail, be put under cultivation. It is, therefore, safe to assume that the immense landed property of the Company, as a body, in connection with valuable town sites and water-powers, will ultimately be worth much more than the entire cost of the Railroad. It should be borne in mind that where there are now only prairies, subject to continual burnings, trees will spring up and flourish hereafter when these lands become protected by civilization. The growth of trees on the Pacific slope is very rapid.

SNOW.

There is evidence enough to show that the line of road, on the general route herein described, will, in ordinary winters, be much less encumbered with snow where it crosses the mountains than are the passes at more southerly points, which are much more elevated above the sea. The difference of five to six degrees of latitude is more than compensated by the reduced elevation above the sea level and the climatic effect of the warm ocean currents from the equator, already referred to, ameliorating the seasons from the Pacific to the Rocky Mountains. An examination of the profile of the Union Pacific and Central Pacific lines between Omaha, on the Missouri river, and Sacramento, California, a distance of 1775 miles, shows that there are four main summits; Sherman summit, on the Black hills, about 550 miles from Omaha, 8235 feet above the sea; one on the Rocky Mountains, at Aspen summit, about 935 miles from Omaha, 7463 feet; one at Humboldt mountain, about 1245 miles from Omaha, 6076 feet, and another on the Sierra Nevada, only 105 miles from the western terminus at Sacramento, 7062 feet; whilst from a point west of Cheyenne, 520 miles from Omaha, to Wasatch, 970 miles from Omaha, a continuuous length of 450 miles, every portion of the graded road is more than 6000 feat above the sea; being about one thousand feet on this long distance, higher than the highest summit grade on the Northern Pacific Railroad route, whilst for the corresponding distance on the Northern Pacific route, the average elevation is under 3000 feet, or 3000 feet less. The highest summit on the Northern Pacific line is about *three thousand feet* lower than the Sherman summit on the Pacific line.

On the Union Pacific road the profile also shows that for nine hundred continuous miles, from Sidney westward, the road has an average height of over 5000 feet, and the lowest spot on that distance is more than 4000 feet above the sea; whereas on the Northern route only about sixty miles, at most, are as high as 4000 feet; and the corresponding distance of nine hundred miles, extending from the mouth of the Yellowstone to the valley of Clarke's river, is on average about 3000 feet lower than the Union Pacific line. Then, allowing that 1000 feet of elevation causes a decrease of temperature of three degrees, this would make a difference of nine degrees. There is, therefore, a substantial reason for the circumstance, now well authenticated, that the snows on the Northern route are much less troublesome than they are on the Union Pacific and Central Pacific route. At the same time it should not be claimed that there will be no trouble from snow on the Northern line. There are places where snow falls to the depth of several feet; and at the higher passes to the north of the estimated line the snows are much deeper than at the Deer Lodge Pass; and also on the alternate line referred to, across the Coeur d'Alene range, the snow falls to a considerable depth. And this is likewise the case on the indicated cut-off routes from the Columbia river, by the way of the Yakima river valley, etc., through the passes of the Cascade range, to Puget Sound. The impression I would wish to create is this: That a line can be so located between the valley of the Missouri and the mouth of the Columbia river, and to Puget Sound, which for the greater portion of the distance will not encounter any serious trouble from snow; and that in the passage of the Belt range, between the Yellowstone and the upper Missouri, in case that that route should be chosen, the crossing of the Rocky Mountains at Deer Lodge Pass, and of the Portage between the Cowlitz river and Puget Sound, no greater obstacles from snow are likely to be met with than have already been encountered and overcome on roads in the New England States and in the State of New York. It is the general impression in the States, an impression entirely natural, that the farther we go to the north the deeper the snow; but on this line the modifying and controlling influences of the mild climate which pervades the Pacific slope and the interior along this latitude, combined with the greatly reduced elevation of the range of country to be occupied, and the low summits of the back-bone mountains to be passed, tend to confirm the favorable statements of intelligent gentlemen who have long been familiar with the regions in question. I cannot hesitate, therefore, whilst fully appreciating the importance of the matter, to present to you confidently these views in relation to the probable snow difficulties along the proposed route of the Northern Pacific Railroad.

GRADES.

The grades on the route across through the State of Minnesota and Territory of Dakota to the Missouri river will not be materially dissimilar to those on the other finished railroads south of it, passing from Chicago to Sioux City, Council Bluffs, etc.; namely, undulating within the general limit of about forty feet per mile, although it may be deemed advisable, at a few points for short distances, to run to a maximum of one foot per hundred, or fifty-three feet per mile. There is sufficient knowledge of this portion of the route to warrant this assumption. And beyond the Missouri, along the valley of the Yellowstone, to near the Bozeman pass, there is no known reason for assuming any higher limits. In passing Bozeman summit of the Belt range, and in going up the eastern side of the Rocky Mountains, it may be found advisable to adopt a somewhat higher gradient for a few miles in overcoming those summits. This, however, can only be finally determined after careful surveys.

The highest ground encountered between Lake Superior and the Missouri river, at the mouth of the Yellowstone, is only 2300 feet above the sea, the low summit of the Rocky Mountains is but little over 5000 feet, and the Bozeman Pass through the Belt range is assumed to be about 500 feet lower. The height of the country upon which the line is traced, and upon which my estimate of cost is based, may be approximately stated thus, beginning at Lake Superior, going westward :—

	Miles.	Average Height above the Sea.
To Dakota valley,	300	1200 feet.
Yellowstone river,	300	2200 "
Along Yellowstone,	400	2500 "
Flathead valley,	300	3500
Lewis or Snake river,	200	3000
Puget Sound,	500	400
Lake Superior to Puget Sound *via* Portland,	2000	

Compare this with the profiles of the finished line of the Union and Central Pacific Roads. Properly, the comparison should be made from Chicago—the eastern water terminus on Lake Michigan, of the Omaha line. There are on that route, approximately, as follows:—

From Chicago.	Miles.	Average Height above the Sea.
To Omaha,	500	1000 feet.
Near Cheyenne,	500	3300 "
Cooper's,	100	7300 "
Promontory Point,	485	6200 "
Humboldt,	406	4750 "
Reno,	130	4000 "
Auburn,	118	4400 "
Sacramento,	36	300 "
San Francisco,	100	50 "
Chicago to San Francisco,	2375	

On the Northern Pacific line there need be but two principal summits, whilst on the other there are four; the lowest of which is about a thousand feet higher than the highest on the northern route. If, therefore, the roads were the same length between the Pacific waters and the great lakes and navigable rivers east of the Rocky Mountains, the advantage would be largely in favor of the northern route; but this actual distance is 375 miles less, and the equated distance for ascents and descents in its favor will be very considerable.

The first fairly navigable water for steamers which is met with on the route from Sacramento eastward is the Missouri river at Omaha, 1775 miles; whereas on the Northern line the road will be along or quite near navigable streams a large portion of the distance on the west side of the Rocky Mountains; and soon after crossing them it will strike navigable waters on the eastern side, on the Upper Missouri and on the Yellowstone. The mountain space between the navigable streams on the two sides of the range is only about 250 miles. It is true that the Union Pacific Road runs along near the Platte river, but, owing to its peculiar character, arising from its breadth and sandy shoals, it is not regarded as a practicable steamboat river in its present condition.

The junction of all main trunk railroads that may be built eastward from the Pacific ocean with the Missouri river, will ultimately prove to be much more important than is yet generally imagined; especially when the valley of that great river becomes more densely populated. An extensive interchange of products will take place there, both eastward and westward from the river along the east and west lines. The immense valleys of the Mississippi and Missouri, all the way from New Orleans to the upper waters, are yet in their infancy of population, of agriculture, of manufactures and of general commerce. The construction of railroads along and across these magnificent valleys will stimulate all of these by the great facilities afforded, and throw a large amount of local traffic upon the east and west railroads.

No main trunk line can enjoy a more advantageous position in this respect than the Northern Pacific Road; and to a very considerable extent this will apply in the case of the crossing of the Dakota, the Mississippi, and the Red river; the productions of the Red river, owing to the extent and wonderful fertility of its valley, will at once create a remunerative trade on the eastern end of the Road. This region, it is well known, is admirably adapted to wheat, and intelligent gentlemen who are familiar with its characteristics, have estimated that there are fifty millions of acres of cultivable land in the Red river valley alone, which may be profitably devoted to the raising of wheat and other cereals. Branch lines will undoubtedly be constructed leading from the Northern Pacific trunk along the St. Louis river, the nearest to Lake Superior, as well as along the Mississippi, the Red river, the Dakotah river, and the Milk river branch of the Missouri. These branches will be met by the Canadian improvements, bringing into this connection the extensive regions along Lake Winnepeg, including the valley of the Assiniboine. A branch from the Missouri, reaching to Milk river and up that stream to the great Saskatchawan valley of Canada, will at no distant day constitute a very important feeder of the Northern Pacific line, opening to that large territory railroad outlets to the navigable waters of the Atlantic coast by way of Lake Superior, and to the Gulf of Mexico by the Missouri and Mississippi rivers. No public project that can be devised can sooner or better develope that vast territory in Canada and bring out its value. The summer isothermal line of 70° reaches to the valley of the Saskatchawan about latitude 51°. The same isothermal passes through Chicago, Cleveland, and Harrisburg in Pennsylvania, on this continent, and through Southern France, Lombardy, and the great wheat growing districts of Southern Russia. These valuable Canadian territories, which for the want of adequate investigation have hitherto been regarded as valueless, are destined within a reasonable period to perform an important part in the progress of agriculture and commerce on this continent. Lines of latitude are no longer the sole guides in estimating the climatic characteristics or producing qualities of regions to the north, since experience has shown that the chilling effect of high latitude so marked on the Atlantic coast is overruled by the influence of the Pacific ocean, and perhaps other causes not yet thoroughly elaborated or understood.

But a few years will be required after the completion of the Northern Pacific trunk line to secure what may be termed local trade and travel sufficieut to sustain the road irrespective entirely of any through business. The Territory of Montana, already yielding more than ten millions of value annually, abounding in elements which must induce a large population, will meet the opening of the road with a very valuable contribution, while there cannot be a doubt that accompanying and following the construction of this line hundreds of settlements will rapidly be made in the valleys of the Missouri and Yellowstone on the east, and in the valleys of Clarke's river and Columbia river on the west of the Rocky Mountains. With respect to the local trade and travel on the road along the lower Columbia, and from the Willamette valley, concentrating at the city of Portland, the population and business are already there, only awaiting the opening of this line to make the western end self-sustaining from the start. I speak confidently on this point from personal observation.

An estimate of items of transportation on a road situated as this is, dependent in part, and to a considerable extent on the settlements which will hereafter be made along the route, would of course be largely conjectural; but if experience on hundreds of roads in the United States which have been extended through comparatively unsettled regions is of any value, there can be no doubt of the result on this line. Its cost will be less per mile than the average cost of railroads in this country; and the cost of working it, owing to the predominence of light grades over the greater portion of the route, and the facilities it possesses most of the way in good supplies of water, wood and coal, will fall below the average.

The position across the continent on the shortest practicable railroad distance between the Pacific ocean and the great lakes of the Atlantic side, points to this line as one of vast importance in a national point of view, the value of which to the Government cannot easily be overrated. The facilities it will afford for the rapid and economical distribution of troops, ammunition and stores, for the numerous forts on the waters of the Missouri and the Yellowstone, and along the valleys of Clarke's river, Columbia river, and on Puget Sound, will constitute an invaluable military arm, and will save millions annually to the public treasury. A moderate estimate of the mere money saving to the country will show not less than three per cent. on the entire cost of the road every year, for present military transportation alone, to forts now in existence; and more forts will be needed to hold the Indian tribes in check as the white settlements shall be annually extended over Indian territory—a matter which is inevitable, and as certain as the ultimate extermination or absorption of the Indian race. The opening of this road will forever settle the question of white supremacy over an area of country covering at least four hundred and fifty thousand square miles; sufficient to make ten States the size of Pennsylvania.

The Northern Pacific Railroad will be a favorite and most convenient route for trade and travel across the ocean from Asia and the Sandwich and other islands of the Pacific; transporting

Asiatic products quickly to the gold regions of Montana, and distributing them along the valleys of the Yellowstone, the Missouri, the Dakota, the Red river, the Mississippi, and the Lake region. It is true that the main terminus will be at Lake Superior; but connections with St. Paul and Chicago are certain to be made which will put this line in direct communication with the entire railroad system of the United States and Canada.

The prevailing winds and ocean currents of the Pacific are such that vessels from Asia, etc., destined even as far south as San Francisco, are compelled to make northing which brings them on the parallel with Puget Sound. This is a very important fact in connection with the Northern Pacific route, inasmuch as it has a permanent practical advantage which a great commercial city located on Puget Sound, or on the Columbia river, will enjoy over any port six or eight hundred miles south along the coast.

I have deferred the presentation in detail of many points, among others the harbor termini on Lake Superior, on the Columbia, and on Puget Sound; although I am possessed of information which will enable me at the proper time to discuss the advantages and disadvantages of the several places proposed.

This special report contains, in brief, the substance of the information obtained during an exploration which occupied the entire months of June, July and August; involving in all over nine thousand miles of travelling, on railroads, in coaches, steamers, wagons, canoes, and on horseback; during which, owing to the remarkable facilities afforded through the aid of the modern conveyances by steam, both on land and water, our party was enabled to explore an extent of territory which in the time of Lewis and Clarke occupied nearly two years.

In conclusion, I would state as the result of these explorations and investigations, after much reflection, and fully appreciating the responsibility devolved upon me as the Engineer selected by you for the duty, that the Northern Pacific Railroad route, with the land grant secured to the Company by the Government, possesses great intrinsic value, and will be, as a whole, a remarkably favorable line in all important respects; a line which, if judiciously located, honestly constructed, and properly administered, will pay within a few years a fair dividend on its cost. I had apprehensions that personal investigations might disclose material or possibly vital errors in some of the anticipations induced by former Reports. The result, however, has been in the other direction; and I am constrained by the facts to present an estimate of cost essentially lower than those previously submitted by the able Chief Engineer, and I offer it confidently as reasonable and reliable.

Very respectfully,

W. MILNOR ROBERTS,

U. S. Civil Engineer.

PHILADELPHIA, September 25, 1869.

(3)

NORTHERN PACIFIC RAILROAD.

PROCEEDINGS

OF A

CONVENTION

HELD AT

EAST SAGINAW, MICHIGAN,

ON THE

23d and 24th November, 1869.

TO CONSIDER THE QUESTION OF

THE SHORTEST AND MOST FEASIBLE ROUTE, AND THE BEST MEANS FOR PROMOTING THE CONSTRUCTION OF THE NORTHERN PACIFIC RAILROAD.

EAST SAGINAW:
Daily Enterprise Office

1869.

A Call for a Northern Pacific and Michigan Short Line Rail Road Convention, at East Saginaw, Michigan, November 23d, 1869.

The undersigned committee, appointed by the Common Council of East Saginaw to issue the call and make arrangements for a convention to be held in this city on the 23d day of November, 1869, at 2 o'clock P. M., respectfully and urgently request all municipal and other corporate bodies on the contemplated line of the Northern Pacific Railroad, from the head of Lake Superior, *via* the Straits of Mackinaw, to the Atlantic, to send delegates to said convention.

And it is earnestly hoped that Boards of Trade, City Councils, county and township authorities, railroad corporations, and other bodies throughout the State of Michigan, will be fully represented; and that the States of Iowa, Minnesota, Wisconsin, New Jersey, New York, Connecticut, Rhode Island, Massachusetts, Maine, New Hampshire, Vermont, and the Province of Canada, will send delegates, especially from the Western States named, and from along the line of the route recently adopted by the Trans-Continental Railroad Convention at Oswego.

The following are among the important questions to be submitted for the consideration of the convention:

1st. Is there any necessity for more railroads across the continent? If so, what routes should be adopted?

2d. Is there any advantage to be secured to the Northern Pacific road by reason of its reaching water communication sixteen hundred miles from the Atlantic, at the head of Lake Superior?

3d. Comparative cost of transportation by rail and by water; time being considered money.

4th. Puget Sound as a western and Portland as an eastern terminus of the Northern Pacific Railroad.

5th. The route by the Straits of Mackinaw; is it the shortest and most practicable? Will it open and develop a region of country rich in soil, timber and minerals, and by reason of which, will it secure a vast amount of local traffic?

6th. Do the Central and Union Pacific Railroads supply such advantages to the country as to render it inexpedient for the Government to encourage the building of a Northern and Southern Pacific Railroad?

7th. Should the Government, in the present state of its finances, make further grants to railroads?

8th. The character of the country on the line of the Northern Pacific.

9th The valleys of the Upper Mississippi and the Red River of the North.

10th. The western slope of the Rocky Mountains to the harbor of Puget Sound.

11th. The shortest line between China and Europe.

12th. The railroads of the United States; do they tend to settle, develop and enrich the country?

13th. Would the granting of subsidies to two more Pacific railroads retar l or expedite the liquidation of the national debt?

14th. The railroads of Michigan as the shortest connecting links between the Northern Pacific Railroads and the roads from the East and South

15th. The duty of the Government in reference to the future disposition of the public domain.

16th. The present and prospective value of the agricultural, mineral, saline and lumber products of Northern Michigan

17th. The future of the Northwest.

It may be observed that the title given to this convention, at the head of this call, implies that its object is local and not national in character. To this we may reply that the Northern Pacific Railroad, thought to be a great national trans-continental line, is nevertheless local, so far as its route and *termini* are concerned, and it has been observed that the Trans-Continental Railroad Convention, recently held at Oswego, N. Y , was called with the express object of building an air line from Portland, *via* Oswego, Lewiston and the Detroit or St. Clair rivers, to Chicago, there to meet the Union and Northern Pacific Railroads; and, though that convention approved of the line to Chicago, it also approved of continuing the Northern Pacific from the head of Lake Superior, east, *via* the Straits of Mackinaw and the Detroit or St. Clair rivers.

This convention, then, cannot be considered any more local than the one at Oswego; though its chief objects shall be the building of the Northern Pacific Railroad through Michigan, across the Straits, to connect in the Lower Peninsula with all roads running east and south.

We are free to admit that, in view of her seven hundred miles of undeveloped territory, over which such a route would pass, the richness of her soil, the amount of valuable timber, her sixty thousand inhabitants in the Upper Peninsula engaged in developing her vast deposits of copper and iron, and who have no outlet except by water, and on account of her saline and other comparatively undeveloped elements of wealth, Michigan, perhaps, has a more vital interest in the object of the convention than any other single State.

But while this may be true, it is equally true that all the States named in the call are more interested in this than in any other route. This is made evident by a glance at the map, where it will be seen that this line will open up the northern part of Wisconsin, and afford a competing railroad outlet east, other than by Chicago, for wheat and other products of Iowa and Minnesota.

It will also appear evident that, should the Northern Pacific Railroad adopt no other line to the east than around the south end of Lake Michigan, (which is on the line of the Union Pacific,) its traffic would naturally drift to the South, and New York and all the New England States would be injured, and New Jersey to some extent, thereby. It is not the object of the convention to oppose connections between the Northern Pacific Railroad and Chicago, but to secure a route also by the Straits, and thus to guarantee to the Northern and Eastern States what the Government intended by its grant of land—a Northern Pacific Railroad, to run as far as possible through the undeveloped portions of the States and Territories.

By this route, and the connections with Chicago, all railroads running south and east, and all the States on the Atlantic, would receive their just share and interest in a trans-continental railroad built by the liberality of their Government and the energy of its citizens.

To secure these objects, and to discuss the subjects embraced in this circular, the convention is called, and we again earnestly invite the presence of delegates and of all interested citizens from every quarter.

The undersigned also respectfully request all newspapers friendly to the objects of this convention to publish the call for the same. It was the intention to have the call signed by many leading citizens in this and other localities who approve of the convention, but the shortness of time, and the necessity for an early and extended notice, induces the committee to publish the call with their names alone attached.

And acting as we do, in harmony and in accordance with the desire of all favorably interested, we trust that the importance of the convention will not be diminished, or an interest in the same weakened, on account of the call not being more numerously signed.

Hoping that the response of delegates may be equal to the importance of the subject, we are, with high consideration,

JAMES L KETCHAM, Mayor of East Saginaw,
C K. ROBINSON, WM. J. BARTOW,
JOS. A. HOLLON, JEREMIAH FISHER,
GEO. W. PECK, JOHN F. DRIGGS.
GEORGE W. PECK, Secretary.

PROCEEDINGS OF THE CONVENTION.

FIRST DAY.

Pursuant to the preceding call, a Convention of Delegates assembled at Irving Hall, in the City of East Saginaw, at 2 o'clock, P. M., on Tuesday, November 23d, 1869.

The rear of the platform of the hall was covered by an immense map, exhibiting on a large scale the route, etc., etc. It was prepared by Messrs. Goddard, Kirby, Herbert and Eastman, and proved of great value in the deliberations, etc.

The Convention was called to order by Hon. Geo. W. Peck, and organized by appointing Hon. John F. Driggs of East Saginaw, temporary President, and F. H. Rankin, of Flint, temporary Secretary.

Mr. Driggs, on assuming the chair, addressed the Convention as follows:

GENTLEMEN OF THE CONVENTION:

I only express the sentiments of all the citizens of this valley, when I extend to you a cordial and kind welcome to this locality, and I know I express the regrets of our whole community that an unforeseen and severe storm has probably prevented the attendance of many delegates from distant points. We are, however, very much gratified, notwithstanding the inclemency of the weather, to see so many intelligent gentlemen present. I accept the position of temporary Chairman of this Convention, and would suggest the appointment of a Committee on Credentials and Permanent Organization, to report as early as our evening session.

While waiting your further pleasure. I would say that I feel, gentlemen, as though this is not an East Saginaw, or a Bay City, or a Saginaw City, or in fact, a Convention that has reference to any particular locality in this vicinity, but that is a *great national project*; one which, however, if carried out, will benefit certain localities on the line of this great national railway.

We have received letters from many distinguished gentlemen from different parts of the country, which may be read by the Secretary for your information.

We had hoped that we should secure the attendance of the delegation from our own State in Congress, as we look for aid from that quarter to help us build the Northern Pacific Railroad, and especially from the head of the lake down through Northern Michigan, and on further East, so as to connect with other contemplated lines.

I am sorry to say that for want of time we were obliged to defer the Convention so late as to bring us into quite a severe beginning of winter; at the same time could not call it earlier for want of time after the project was once thought of, to spread the invitations sufficiently wide.

You will permit me one suggestion: At the Convention that I recently had the honor to attend at Oswego, N. Y., the Convention was opened and closed with prayer. Whatever may be our individual opinion in regard to this, it is a respect that we owe to the sentiment of the community at large, and in accordance with a time-honored custom. At the request of a large number of delegates present, I will invite Dr. McCarty to open the Convention with prayer.

The Rev. Dr. McCarty thereupon invoked the divine blessing on the deliberations of the Convention.

Hon. Wm. Phelps, of Detroit, moved that a Committe of five be appointed to examine the credentials of Delegates, and report the names of those present who are entitled to seats in the Convention, which motion prevailed, and the Chair appointed as the

COMMITTEE ON CREDENTIALS,

W. H. Craig, of Detroit, J. A. Hubbell, of Houghton, C. K. Robinson, of East Saginaw, Hon. D. H Jerome, of Saginaw City, and Hon. Randolph Strickland of St. Johns.

J. A. Hubbell, of Houghton, moved that a Committee of five be appointed to report the names of Officers for the permanent organization of the Convention, which was carried, and the Chair appointed as the

COMMITTEE ON PERMANENT ORGANIZATION,

Hon. Wm. Phelps, of Detroit, H. M. Fitzhugh, of Bay City,

Hon. Eugene Pringle, of Jackson, Artemas Thayer, of Flint, and G. Morton, of Monroe.

The Chair announced that a number of letters and telegrams relating to the object of the Convention had been received from prominent gentlemen who were unable to attend in person; and with the permission of the meeting they should be read while the Convention was awaiting the reports of its Committees just appointed. The Secretary accordingly read the following:

TELEGRAM FROM THE MAYOR OF GRAND HAVEN.

GRAND HAVEN, Nov. 23, 1869.

TO HON. JAMES L. KETCHAM:

In any settled plan adopted in your Convention relating to the interest of Michigan and the Northern Pacific Railroad, the city of Grand Haven will concur. Local interests requiring my personal attention prevent my being with you.

DWIGHT CUTLER, *Mayor.*

LETTER FROM THE MAYOR OF OSWEGO, N. Y.

OSWEGO, Nov. 18, 1869.

HON. JOHN F. DRIGGS,

Dear Sir:

I very much regret being unable to attend your Convention on the 23d inst. It is an important Convention and I wish you success.

I wish to introduce to you Mr. John A. Barry, Editor of the Daily *Palladium* of this city, with whom I think you became somewhat acquainted when here at our Convention. Mr. Barry is probably the only one of our delegation able to attend. He will represent us in a proper manner, and give a good account of the proceedings in his paper.

Very Truly Yours,

A. S. PAGE.

LETTER FROM JAY COOKE.

PHILADELPHIA, Nov. 19, 1869.

HON. JOHN F. DRIGGS,

East Saginaw.

Your favor of 15th inst is received. I must answer the questions proposed in your call very briefly, if at all. These questions number seventeen. I will take them in their order:

1st. In my opinion there *is* immediate necessity for more railroads across the continent. Two, at least; the Northern and Southern roads, should be begun at once and completed within three or four years. If our Government expects the fealty of these vast regions, it should bind them to the East, and make the inhabitants of the territories and the Pacific States feel that they are a part of the United States, practically as well as in name.

2d. I think the Northern Pacific Road secures a vast advantage by reaching water communication, which at the head of Lake Superior, is as near the Atlantic as Chicago is; but this advantage is only for a portion of the year, and does not apply to light freights or first class travel. It does apply, however, to a vast trade in grain and other articles, and to emigrant travel

3d. I do not consider it worth while to argue the comparative cost of railroad and water transportation. There will be enough for all to do when the vast regions to the West and Northwest of Lake Superior are opened to settlement and cultivation.

4th. If Canada belonged to us, we would soon cross the Sault Ste. Mary, skirt the north shore of Lake Huron, run down the Ottawa to Montreal, thence to Portland, thus making that city with Puget Sound; but as Canada is not ours, and not likely to be soon, Portland, as a terminus, is out of the question, and we cannot enter Canada at the Sault Ste. Mary. The Northern Pacific, after touching the lake at the mouth of Montreal river, would seek the best route to the Straits of Mackinac. Crowning these, it would seek an outlet by the various roads centering at Detroit, Toledo, etc. Of course the road would send a portion of its business, *via* St. Paul, to Chicago, St. Louis, etc.

5th. I am satisfied that, unless there are insurmountable difficulties at the Straits of Mackinac, any railroad seeking an eastern outlet *via* the south shore of Lake Superior and the Peninsula would find it to the advantage of the road to take this route.

6th. I have answered this in my reply to the first question.

7th. I do not think the Government should, at present, make direct grants of money or bonds for any purpose except for the ordinary river and harbor, and other similar objects, and other improvements. But I do think that some plan should be devised whereby these great trans-continental improvements may be at once commenced and rapidly completed, and I am confident that aid can be given in such form as will not increase the public debt one dollar, or jeopardize the interests of the public treasury, and yet will be all-sufficient to secure the capital required. Should such aid be granted, it should only be on condition that the Government should hold a first lien on the road franchises lands and rolling stock. This aid would be effectual if given in the form of an endorsement of guarantee of the principal and interest of the roads, and such plans could be adopted as would provide for the interest from the proceeds of bonds, so that the Government could not possibly be called upon to advance the funds for the interest for several years to come, if at all, for it may safely be assumed that, when finished, these roads will earn the full interest on their cost the first year.

8th, 9th and 10th. I send you a report just made by W. Milner Roberts. This report I do not desire published, as it is *not yet made public*, but it must unmistakably confirm all that has been said heretofore of the glorious character of the country along most of the route, and that tributary to the Northern Pacific Road.

11th. The North Pacific route, *via* Puget Sound, would shorten the distance to China, Japan, etc, over eight hundred miles.

12th. This question, of course, is answered by *all* in the affirmative.

13th. The national debt would be liquidated more rapidly by these enterprises becoming successful. The area of taxation would be greatly enlarged, and all parts of the nation would be invigorated and made stronger.

The remaining questions I must pass by, except a brief reference to the 17th and last. We who were born in the West, and have seen the wondrous progress of the last 40 years, *cannot* be too sanguine in our anticipations of the future glory of the North-West. I dare not trust my pen to foretell the future, as I expect to see it, should I live twenty years longer.

I regret that I have not time to do justice to the whole subject as contained in your questions. I hope the Convention will do what it can to influence Congress to grapple with the vital subject, and to enact such laws as will, while protecting the Government to the uttermost and not increasing the public debt, yet give effective and prompt aid to these enterprises.

Sincerely Yours,

JAY COOKE.

LETTERS FROM SENATOR HOWARD.

DETROIT, Nov. 7, 1869.

MY DEAR SIR: I am very glad to see the call for a Northern Pacific Railroad Convention, to be held at East Saginaw on the 23d inst. The time fixed is so near the commencement of the next session of Congress that I may be prevented from attending, though I should be right glad to be on hand.

The series of questions to be discussed are all extremely important. One more might be added, and I think it should be, viz.: What, if any, distance will be saved in running the road down our Upper Peninsula and across the Straits of Mackinac to the Eastern markets, rather than from the head of Lake Superior to Chicago? There is some dispute about this, and the question ought to be carefully and fully considered. My own idea is that the Mackinac route is considerably shorter. It is important, however, that the thing should be clearly verified, and not left to conjecture. You will at once see the importance of the question.

I might add another query—the practicability of running a train of cars across the Straits of Mackinac; but, as that is a pure question of civil engineering to be determined by scientific men, I would not add it. Mr. Johnson, the engineer of the N. P. R. R. Co., has often assured me that the thing is entirely practicable, and I believe him. He is a very sensible, well-instructed man. I am looking with great interest for his report to his Company of the surveys and explorations of the Company the past year.

As to a cash subsidy for this railroad, from Congress, I candidly think the idea had better be given up. If we grant a subsidy in bonds or indorsements to one Company, we must to another, and the railroad (Kansas Pacific) on the 35th degree and the famous Fremont line on the 32d degree—and Heaven knows what other lines of Pacific railroads—will assert their claims to the like aid from Government. The treasury can't stand this whole brood of suckers; their hunger would exhaust a world of subsidies. And this, you may rely, is the feeling. No more cash subsidies to railroads is, you may take it for granted, the sentiment of the people and of Congress.

You remember the omnibus bill reported by a majority of my committee in the Senate. Well, that bill is as "dead as a smelt." There is no chance of its passing whatever. It is dead and buried. The N. P. R. R. was provided for in that bill. I told Mr. Rice, Mr. Canfield, Mr Smith, and others interested in the Northern Pacific, that there was no chance of its passage; no, not at all.

But the Northern Pacific Company need not despair. They have their destinies in their own hands, and, with careful management, *can build the road* and make money. Their charter grants them (substantially) a strip of land 40 miles wide, across the continent from the western boundary of Minnesota, and 20 miles wide east of that line—land enough for an empire—more land, I believe, than is embraced in the State of New York—the most of which is rich and productive, covered with timber, full of other resources, and traversed by numerous navigable streams, facilities which, as everybody knows, do not exist upon the U. P. R. R, the Central P. R. R., of California, the Kansas P. R R, or the railroad on the 32d degree. Again, the U. P. and the Central (forming one line) pass over mountain ranges 8 000 feet above the level of the sea, and have to be "shedded" against snow slides, while there is no point where the Northern Pacific will pass a height more than 3,500 feet above the level of the sea This advantage in favor of the Northern route is inestimable.

As to the obstruction from snow along the Northern line, the testimony is conclusive that it will not be greater than on the roads through New England and Pennsylvania. The further you go West, on the same parallel of latitude, the milder the temperature becomes. This is physically proved by years of careful observation on the part of scientific men who have for long years wintered and summered in those regions. Why, Mr. Flanders, the recent delegate from Washington Territory, who resides on latitude between 46 and 47 in that Territory, told me, last session, that the peach, plum, cherry, Indian corn, and other products of the 41st to the 42d degree, all grow and mature in that climate as finely as in any other part of the world. I don't foresee much obstruction from snow and ice on that line—very little.

From Shanghai, in China, to Chicago, that route will be 500 miles shorter than by way of San Francisco. What a saving in transportation is here! Again, Puget Sound, the western terminus of the road, is the best harbor in the world. Any number of ships can take shelter there, no matter what their draft of water. And, with the exception of San Francisco, it is the only harbor worthy of the name from San Diego, the southern extremity of California, to Sitka, in Alaska.

But time would fail me to go into all the considerations to show the importance of this route to the Pacific, and I must close by expressing the conviction that within a few years it will be apparent to the whole world that it is the best route to the Pacific and to the East Indies across the continent.

The opinions which the Convention may enunciate will have great weight, and will, for good or evil, bear powerfully upon the pecuniary interests of that Company. They should be well considered, therefore, and founded upon firm data.

It is not to be denied that the States more immediately interested in this great work are Northern States—all New England, New York. Michigan, Wisconsin, Illinois, Minnesota, Iowa, Nebraska. The Territories of Dakotah, Montana, Idaho, Washington, all have a deep interest in the enterprise These communities are at least something in the scale of commerce and political power, and their voice, in both respects, is potent.

Yours truly,

HON. JOHN F. DRIGGS. J. M. HOWARD.

DETROIT, Nov. 21, 1869.

DEAR SIR: Your answer to my former letter came early to hand. What I said about cash subsidies to railroads was advanced simply as my own opinion as to the future action of Congress, not as indicating what my own vote would be in a proper case. I ever thought, and still think, that the Northern Pacific Railroad is as deserving of such a subsidy as any line that has received it, and have labored hard, heretofore, to secure it for that road. If I could have my way in the matter, I certainly should extend the credit of the Government to that work, while I am not prepared to say the same of any other line that has been recently projected. I mean the lines named (with that line) in the Senate "Omnibus bill" of last session, because I see that the Northern route is to be, one day. the great route to the Pacific and the East Indies, and to give the people of this country a decided advantage even of the Suez Canal.

You are, of course, at liberty to make such use of my letters as you think best.

I should be very glad to attend the Convention, but as my children are all from home, and I am preparing to start for Washington, I find it extremely difficult to do so.

Very truly yours,
J. M. HOWARD.

HON. JOHN F. DRIGGS.

LETTER FROM GOVERNOR BALDWIN.

EXECUTIVE OFFICE,
DETROIT, November 22, 1869.

HON. JOHN F. DRIGGS, East Saginaw.

DEAR SIR: I am in receipt of your favor of Nov. 19, inviting me to attend a Railroad Convention at East Saginaw, on Tuesday, the 23d inst I should certainly endeavor to be with you on that occasion but for the meeting of the various State Boards, at Lansing, on the morning of the 24th, which requires my presence at the capital at that time.

No other class of public improvements has done so much for the growth, advancement and prosperity of the country as the construction of railroads. I am not, however, one of those who advocate the building of railroads without regard to the character and value of the country from which they are to derive their support.

The natural advantages of Michigan are unsurpassed by those of any other State. I have for many years felt the importance of, and have advocated the construction of, a railroad through the center of the Lower Peninsula to the Straits of Mackinac, with east and west branches, from the vicinity of Houghton Lake to the Grand Traverse country on the west, and Thunder Bay on the east. Such a road would traverse a country of unsurpassed fertility of soil, and covered with a variety of timber of untold value. Yet the very fact of its being heavily timbered has thus far retarded such improvements in that section of the State, because of the somewhat increased cost of construction, and of the greater difficulty in bringing immediately into a state of cultivation land covered with a heavy growth of wood. Other causes have also operated to keep back the construction of railroads in the northern portion of our State; these, however, are now being done away.

At the present time railroads are being built in almost every portion of the State, some of which are stretching out their arms towards the north.

The southern portion of the Lower Peninsula has, for a long time, had its three great lines running east and west—the Michigan Southern, the Michigan Central, and the Detroit & Milwaukee. From each of these are lines diverging both north and south to a reater or less extent.

Without pretending to name the many roads in operation, in process of construction, or projected, tending to the development of the State, there are some great trunk lines partially finished, in operation, and in progress of completion, which have a direct bearing upon the subject, to consider which the Convention has been specially called

The Flint & Pere Marquette, with its initial or starting point at Holly, on the Detroit & Milwaukee, on the eastern side of the State, already opened to Midland and to Bay City, and soon to be completed to Pere Marquette, or some other point on Lake Michigan —the Jackson, Lansing & Saginaw road, with its starting point at Jackson, in the center, already in operation to Saginaw and Wenona, and taking measures to continue its line to Houghton Lake and beyond—the Grand Rapids & Indiana Railroad, on the west, from the southern line of the State at Sturgis, already opened from Grand Rapids forty miles north, and steadily progressing towards Grand Traverse or the Straits of Mackinaw—these three great and most valuable lines of roads are steadily going forward.

From the vicinity of Houghton Lake north to the Straits of Mackinaw there should be but one road. Could all interests be united, and, with this union of strength, a single road be constructed from Houghton Lake to the Straits, with lateral branches to Lake Huron on the east and Traverse Bay on the west, it could not fail to result in the rapid development of the northern part of the State; population would flow in with a rapidity hitherto unknown, enriching the State and at the same time furnishing a large local traffic for the road.

So far I have alluded to the development of the southern peninsula only, but we should not overlook the resources of the northern peninsula, rich, almost beyond calculation, in its mineral wealth of iron and copper. Most wisely, in my judgment, the Constitution prohibits the State from being interested in any work of internal improvement, or carrying on any such work, except in the expenditure of grants to the State of lands or other property. It is not necessary, to secure the construction of these roads, that the State should undertake the work. Congress has made large appriations of lands for this purpose. These lands, already valuable, will greatly appreciate by the construction of the roads.

Our country is yet in its infancy. Population is flowing to us, from every part of the world, with wonderful rapidity. One great line of railway already unites the Atlantic with the Pacific. Another shorter and better line, the Northern Pacific, is also to be constructed. It is simply a question of time, and no people can be more deeply interested in the adoption of this line than the citizens of Michigan It is not necessary to oppose connections between the Northern Pacific road and Chicago, but, whatever may be its connection with other lines around the south end of Lake Michigan, it should be the policy of the Government to make the line, as far as possible, through the undeveloped portions of the country, and especially so if that country is, as in this case, of valuable character and capable of sustaining a large and dense population; and still further, when such line would shorten the distance to the large manufacturing States of New England, and to Europe from the sea-board.

Every consideration would seem to favor the construction of a road through Northern Michigan, and making that road a part of the line of the great Northern Pacific.

I trust that the deliberations of the convention will tend to hasten the construction of the road through both peninsulas of our noble commonwealth, and to further the undertaking and completion of the great practical northern line to the Pacific coast.

Sincerely regretting my inability to participate with you in the proceedings of the convention, I am,

Very respectfully yours,
H P. BALDWIN.

LETTER FROM HON. JOHN A. POOR.

PORTLAND, Nov. 18, 1869.

HON. JAS. L. KETCHAM, Mayor of East Saginaw, and others.

GENTLEMEN: I have the honor to acknowledge the receipt of your invitation to attend a convention, at your city, on the 23d inst., of those friendly to the contemplated line of the Northern Pacific Railroad "from the head of Lake Superior, by the way of the Straits of Mackinac" and across the State of Michigan, to a point of connection with the line of the proposed trans-continental railway from the St. Clair river to Portland, Halifax, and the most eastern shore of Newfoundland, and regret that prior indispensable engagements compel me to forego the pleasure of a visit, at this time, to your interesting section of the country.

In this connection it gives me pleasure to inform you that a section of the European and North American Railway, 80 miles in length, between the River St. Johns, at St. Johns city, and the River St. Croix, the eastern boundary of the United States, is to be formally opened for business on the 1st of December—at which opening I hope to be present—leaving but 56 miles of unfinished line, now in rapid progress, between Bangor and St. Johns city, the entire line from Halifax to Bangor being all in progress, to be opened through, 600 miles from Portland to Halifax, during the year 1870.

Portland and Saginaw are upon the same parallel of latitude, and the deflection from an air line to Portland, which you are compelled to make to pass Lake Huron, is scarcely sufficient to be taken into account in projecting a great line of railway to the sea-board.

We all recognize Chicago as the present great commercial point in the Northwest, but any one must perceive, on reference to a map, the extraordinary advantages for a great city at the head of

Lake Superior, and it is apparent to any one that, from the head of Lake Superior to the foot of Lake Huron, the shortest practicable route will be found by the Straits of Mackinac, resorting to a ferry to connect the two sections of such a line.

The interesting manner in which the claims of the enterprise were presented to the Oswego convention by its President, the Hon. John F. Driggs, for many years your able and honored Representative in Congress, with the aid of the great map present at that convention, produced a decided and profound impression, and while the call for the Oswego convention looked mainly for a line from Portland to Chicago, we could not fail to see that the great Northwest, lying far above Chicago, was ultimately to be reached from the Atlantic sea-board, by the direct line contemplated by your call, to the head of Lake Superior by the Straits of Mackinac.

I will not attempt to describe the present value or the mighty future of the vast interior, which will naturally connect itself by rail with the head of Lake Superior, lying upon both sides of the boundary line. But there is a vast, unoccupied region, destined in the future to sustain as dense a population as that which now inhabits the same latitudes on the Eastern Continent, in Northern Europe and Asia, and the testimony shows that wheat can be raised as far north as the 60th degree of latitude, and Indian corn over a portion of this region of the vast and well watered table lands west and northwest of Lake Superior.

I had occasion, in a memorial to the Legislature of Canada, in 1854, to say "that from the shore of the Atlantic ocean, at Portland, to the head waters of the Missouri river, every portion of the country presents the greatest possible attractions to invite immigration; that a mild climate, healthy at all seasons of the year, a soil of abundant fertility for the production of food, with rich mines and luxuriant forests, are here found for the hundreds of miles beyond the head of Lake Superior, if not to the base of the Rocky Mountains.

"If the waters of these mighty rivers could be turned into Lake Superior, and be at the same time made navigable to the sea, we might form some idea of the value and importance of a railroad from Portland to the head of Lake Superior, and to the region of country beyond it.

"The distance from the great bend of the Missouri, by railway, to the Atlantic ocean, at Portland, would be hundreds of miles less than by following the Mississippi and Missouri rivers to their mouth at New Orleans."

Engaged as you are in the carrying out of a most important section of a great national and international work, I cannot fail to express my hearty sympathy in your labors, and my wishes for its abundant success. Very respectfully,

Your obedient servant,

JOHN A. POORE.

LETTER FROM THE VICE-PRESIDENT OF THE R., W. & O. R. R.

ROME, WATERTOWN & OGDENSBURG RAILROAD,
VICE-PRESIDENT'S OFFICE,
NEW YORK, Nov. 16th, 1869.

HON. JOHN F. DRIGGS,

DEAR SIR: I received, a few days since, a circular letter, under your frank, inviting me to attend a Railroad Convention to be held at East Saginaw on the 23d.

I have delayed answering till now, that I might be able to state definitely whether it would be possible for me to be present, and am now obliged to say that I must forego that pleasure. From the number and variety of topics to be presented for consideration, I conclude that the meeting is for the discussion of railroad extension generally; doubtless, however, intended to have a special reference to a few particular projects named in the circular.

The general question of railroad extension is one in which I feel a deep interest. Taking the large view, and not being confined to those projects that are in, or tending in, the direction of the road with which I am connected, I have had an opportunity to observe some very marked instances of roads which have been undertaken, and finally completed under great difficulties, into new sections, where the construction of a road, at all, was regarded by many as impracticable, and by others as premature, but I have lived to see the immense benefits, at first confined to the country traversed by the road, in a very few years reflected back again to the road, and the capital employed in its construction returned *with good interest.*

As far as I am informed, I am inclined to believe the route in which you are, I think, particularly interested, is not dissimilar from the instances named, with this advantage in its favor: That to local earnings, which must be large, derived from the whole of Northern Michigan, now only needing such an inlet and outlet to make it equal to any other section of the State,—to this will be added a large through traffic from the Superior region, and especially from the head of that lake and proximate to it, already becoming the greatest wheat producing region in the West, and also from the connections that will be made with other roads, and especially with the "Northern Pacific," which may be counted upon, with great certainty, to be constructed within a very few years. For myself, I regard the inducements to the capitalist, and to the General Government to undertake this work, as far greater, in every point of view, than was the case in regard to the Pacific road, already built, when it was projected.

Great and extended as our American railway system has already become, it is but the beginning of that which we shall behold in the near future. We shall also witness equally great changes in the adaptation not only of the roads themselves, but of the equipment and management of the same, tending to a more general use of this system. It is now rapidly overcoming, and soon to defy, competition with any and all other means of transportation of property, as it has already done in the conveyance of persons.

It cannot be doubted, so far as the area of land is concerned, as well as the perfect adaptation of soil and climate, that we are now able to supply the world with bread and the other elements of food dependant on the soil for production. The only obstacle that can be discerned to our doing so is the one contingent upon the cost of transportation from the field to the points of distribution. That this is to be accomplished by the roads built, and to be built, cannot be doubted Every other question affecting the future growth, greatness and prosperity of our great "Empire of the West" sinks into utter insignificance as compared with this, which applies equally to the districts in which cotton, sugar, tobacco and rice, the great *American staples*, are produced.

I must not ask your attention longer to what might be added upon a question of such interest; and again regretting that I cannot meet yourself and other railroad men at the convention, which, I trust, may result in accomplishing all that is expected,

I am very truly yours, etc.,
MARCELLUS MASSEY.

LETTER FROM HON. JOHN NEAL.

PORTLAND, MAINE, 13th Nov., 1869.

TO JAMES L. KETCHAM, ESQ., Mayor of East Saginaw, and his Associates of the Common Council.

GENTLEMEN: I have delayed answering your invitation till to-day, hoping so to arrange my business that I could be with you and assist in your deliberations; but as I find now that I cannot do this, allow me to send you a few words of congratulation and encouragement.

On looking at the map, to which you call our attention, I find unquestionable indications of that same providential foresight which buried all the treasures of earth at the beginning—all our precious metals, our gems, our anthracite, our bituminous coal and our peat, as well as our iron, our lead, our zinc and our marbles—at a depth exactly corresponding to our necessities, our growth and our energy, obliging us to inquire after them and to *dig for* them, just as our agricultural riches have to be dug for, and are always found just *where* and *when* they are most needed.

Just so has it ever been, and is now, with our magnificent railroad enterprises. When they are wanted, they are always found. The signs are all along our pathway, and are always *cropping out*, like our mineral treasures, only to be overlooked till they are wanted. And what is more, when we go to the map, after due consideration, we find our boldest undertakings foreshadowed there by the very lay of the land, so that we begin to wonder why they have never been thought of before.

The mountains and rivers testify for your encouragement, just as in Maryland and Pennsylvania; iron and coal, and lime for the flux, with sand for the furnace, are found lying side by side, for hundreds of miles upon a stretch, overhung by a wilderness of growing charcoal, and all to be had, if not for the asking, certainly for the digging and chopping, the burning and blasting.

Under this view, the map you refer to is of itself a prophecy. You find a way left open to you from the beginning of the world; a pathway for nations over seas and continents.

Ten years ago the road you have in view was not *wanted.* Had your magnificent scheme been projected then, it would have been a dead failure, not for a season, only, but perhaps for generations. And why? Because it was not then *needed,* and being attempted too early, the failure would have discouraged your posterity—like the enterprise of Darien; it would have been postponed indefinitely, as being both visionary and preposterous.

Fifty years ago the golden treasure-house of California would not have been reached, though *dug for.* The time had not arrived. But now, when all the long-buried riches of our earth are honestly wanted, and whole communities are bestirring themselves, they are found cropping out everywhere along the highways of commerce, and always just where they are most needed and are most faithfully and steadfastly inquired after and *dug for.*

Just so will it be with your projected railway from Puget Sound to the harbor of Portland, Maine It is now *wanted,* and the want is felt, from the Atlantic to the Pacific, as a great necessity, alike unappeasable and active. Can there be a doubt, then, of its triumphant and early success? Look at the map once more, and see if you do not find there all the encouragement you need. I have called it a prophecy. It is more; it is a poem, a magnificent epic, where the machinery of two worlds must soon be employed. You may see all this, if so disposed, in the evident provision therein made for the fulfilment of the vast undertaking you have now entered upon.

But you, gentlemen, are business men; you have come together for business—you *mean business;* and I have only to add that, in my judgment, the future is full of promise for you, and that the sooner you sink your shafts, and the sooner your furnaces are in blast for smelting the ore committed to your charge, the sooner you will understand what the Builder of the Universe had in view when He laid out this magnificent highway for you, through mountain ranges and over rivers—over seas and continents, rather—for such it will be in time, and long before this generation has been gathered to its fathers.

Wishing you the success you so well deserve, and I could not well say more, I am, gentlemen, your friend and fellow-laborer,

JOHN NEAL.

LETTER FROM HON. T. W. FERRY.

GRAND HAVEN, Nov. 17, 1869.

GENTLEMEN: I with pleasure acknowledge the receipt of a printed call and letter inviting attention to, and presence at, a convention to be held at East Saginaw on the 23d inst., looking to a speedy completion of the Northern Pacific and Michigan Short Line Railroad. Unavoidable public duties, I regret to say, will prevent the possibility of my being present.

The call has been published in the papers of this city, and the attention of the Common Council thereof invited to the subject

I need not say that the project meets my cordial approval, for no one identified with the growth and prosperity of Michigan could lend other than hearty co-operation.

Trusting that the convention, in the consideration of the vast interests involved, may arrive at some possible plan by which it will be made to appear that the trans-continental railroad, *via* the Northern Pacific and Straits of Mackinac route, is not only the shortest one from the Pacific to the Atlantic sea-boards, but a route thoroughly practicable, I join you in lively hopes for the completion of this great national highway.

I have the honor to be, very respectfully,

Your obedient servant,

T. W. FERRY.

HON. JAMES L. KETCHAM, *Mayor*, and others.

LETTER FROM L. KENDRICK, ESQ.

DRYDEN, LAPEER CO., MICH.,
Nov. 28th, 1869.

HON. JOHN F. DRIGGS, East Saginaw, Mich.

DEAR SIR: Your note of invitation, and circular, is received, and contents noted.

The object of the convention, as set forth in your circular, is one of vital importance to the people of this State, and more particularly to those of the Upper Peninsula and that portion of the State you had the honor to represent in the Congress of the United States.

I have conversed with quite a number of the leading men in the eastern part of our county, and their voice is as that of one man. They regard the enterprise as second to no one of its kind in the nation, and one that must, from the very nature of circumstances, be ultimately a success.

One thing is morally certain—that the Saginaw Valley will be on the line of the proposed road, and whether Detroit or Port Huron should be the next point of importance in the State touched by this great national artery, the result will be about the same, so far as benefiting Northern Michigan Still the people in this locality would be more directly interested in a road that would make Detroit one of its objective points.

I regret very much that circumstances beyond my control forbid my being present at the convention.

We shall look forward, with a great deal of interest, to the result of this very important gathering in your city, and with a hope based upon the best of reasons, shall look forward to the ultimate and final triumph of all you propose.

I am, dear sir,
Yours, very respectfully,
L. KENDRICK.

LETTER FROM HON. WM. E. DICKINSON.

MICHIGAN EXCHANGE,
DETROIT, Nov. 9th, 1869.

HON. JOHN F. DRIGGS, East Saginaw.

DEAR SIR: As I passed through Milwaukee yesterday P. M., I noticed that it was intended to call a Railroad Convention in your city in regard to railroads in Upper Michigan. Will you please send a circular of the proposed meeting to Robert H. Larnborn, Esq., Secretary Lake Superior & Mississippi Railroad, Fifth and Library streets, Philadelphia, Pa.

I shall try and be present myself.

With kindest regards to yourself,

Truly yours,
WM. E. DICKINSON,
54 William street,
(Room 26,)
New York.

LETTER FROM EDGAR CONKLING, ESQ.

DETROIT, MICHIGAN, Nov. 15, 1869.

HON. JOHN F. DRIGGS:

DEAR SIR: I am now on my way to Sheboygan, Michigan, where I shall be for the present and until I can build at Mackinaw City the coming spring. I went to Ohio in 1829, and to Cincinnati in 1841, and now come to your State to reside for the balance of my life, and propose developing my property at the Straits of Mackinaw, of some 35,000 acres, in accordance with the plan set forth in my pamphlet sent you. But for the lateness of the season, I would gladly attend your coming Railroad Convention, on the 23d, at Saginaw, of great interest to your State and whole country.

I began to urge the construction of railroads to and from the Straits, on either side, in all directions, in 1853, and then secured Old Mackinaw, foreseeing the day would come when there would be an important union of rail and water lines of commerce. I have fully posted myself in regard to all that country, and I challenge the refutation of my assertion that there is no point on the northern lakes surrounded with so many elements of manufacturing, and that its agricultural capabilities are greatly underrated, and that the construction of railroads through that section of country to Superior City will develop more latent wealth, in a few years, than that of the settled portion of the southern peninsula of Michigan. No such agricultural and mineral country, with such cheap facilities of transportation for distribution, while on the high road of the most important line between Europe and Asia, can fail to become a populous and wealthy country—much more so than other sections now being settled, much less favored with such elements and proximity to market.

The Northern Pacific Railroad will need all the outlets possible to get, whether by water from Superior City, or by railroad *via* St. Paul and Chicago, or Detroit *via* Mackinaw. No selfish ideas should be allowed in this great national highway. All sections of the country contributing to its aid should equally share in its benefits. The short time limited for its construction suggests the greatest harmony in prosecuting the work and in securing needed legislation. For one, I shall seek to present Northern Michigan to the public in a manner that will invite greater attention to its many superior advantages for manufacturing and commerce.

Yours respectfully,
EDGAR CONKLING.

LETTER FROM HENRY STEPHENS, ESQ.

ALMONT, Nov. 20, 1869.

MR. A. P. BREWER, East Saginaw.

DEAR SIR: I understand there is to be a Railroad Convention at your city on the 22d. I do not think any of our people will be there, and therefore have been requested to write some one who will take the trouble to inform those interested that we, the towns of Almont and Dryden, wish to be on the route adopted to reach the Michigan Air Line Railroad, as we have been informed it is one of the objects of the convention to take measures to organize a company to make such connection, and we would suggest if the road goes to Lapeer, that as Armada is the most northern point on the Michigan Air Line, and the topography of the country between Lapeer and Armada is very level, and no streams to cross, of any size, that this route, being direct, is no doubt the cheapest to build, and can and will raise the greatest amount of aid. I am confident that Almont and Dryden would by vote and subscription raise $100,000 to $150,000, and we want to have the chance to do it. Try and advocate our claims, and you will have our thanks.

Perhaps the best way would be to place this letter in the hands of those who are active and influential in this enterprise, and you can vouch for the writer, I trust, and can also give any information as to feasibility of route.

Very truly yours,
HENRY STEPHENS.

At the conclusion of the reading of the letters, the President said:

If it is the pleasure of the Convention, I will occupy some of its time now in reading a paper that I have prepared myself, more especially with reference to the wealth and probable advantages to be secured to a line in the northern portion of our State. I have prepared the paper at the request of gentlemen of the committee, and I do not wish to intrude upon the Convention now, unless it is their pleasure.

A motion was made and carried that the President proceed now with his report, as proposed. Mr. Driggs then read the following:

PAPER READ BY MR. DRIGGS.

Gentlemen of the Convention: Within the last few years we have witnessed the marvelously rapid construction of one railroad line from ocean to ocean, across the center of the United States. This road was commenced during the rebellion, was prosecuted through a terrible war for its suppression, and completed soon after its close; an achievement, under the circumstances, which no other nation ever equalled; a triumph, within the time, of engineering, of labor, and perseverance, to which the history of the world furnishes no parallel.

The *present* needs two more Pacific railroads, a Northern and a Southern road; and the rapidly approaching future *will have them.* What has been done, under more favorable circumstances, certainly can be done again.

The Government made very large grants of land to the Union and Central roads, and also to the Northern Pacific—the charter to which grants forty sections to the mile from the western boundary of Minnesota, and twenty sections east of that line to the head of Lake Superior. In a letter from the Hon. J. M. Howard, U. S. Senator, he computes this grant to be equal in extent to the State of New York; being a strip across the continent nearly forty miles wide. The greater portion of this vast area of land is unsurpassed in richness of soil, timber and minerals, and highly favored by the even mildness of its climate; yet, without access by means of a railroad, is comparatively without value. Most of it would remain undisposed of at $1 25 per acre, and unoccupied under the homestead or pre-emption laws for many years in the future. But build the road, and all the arable lands within the limits, and other portions valuable for minerals and timber, would be rapidly taken up by actual settlers, and considered cheap at ten times their present value.

The land grant to the Union and Central Pacific roads is believed to be far less valuable than the Northern Pacific grant, by reason of the extended barren plains, the wider mountain ranges of much greater altitude, the deeper winter snows, and the inferiority and severity of the climate on the line over which the roads pass. It cannot be denied, however, that these roads are (for the immediate present) compensated for this deficiency by the fact that their western terminus is at the present populous Pacific metropolis, San Francisco, and that they pass through many intermediate towns and settlements, thus securing an immediate large amount of local traffic and travel. Portions of their lands, however, since the roads are built, are very valuable, while before the consummation of the great work they were entirely insufficient to secure the building of the roads, and might have remained so for fifty years to come

A Government subsidy, then, was necessary to secure the accomplishment of the work. The aid was given, and now the iron horse steams his way in seven days (one-eighth of the distance around the globe) from ocean to ocean, across a continent, over rivers and streams, through valleys and gorges, and over mountain tops eight thousand feet above the level of the sea. If these roads were a necessity, and if the Government acted wisely in extending its aid beyond that afforded by the land grant to the Central and Union Pacific, may it not be wise, also, to extend it to the Northern Pacific, and even to a Southern Pacific road?

From Shanghai to London, *via* Puget Sound, Lake Superior, Mackinaw and Portland, it is considerably nearer than *via* San Francisco and New York, with a much lighter grade; the highest point reached by the Northern Pacific road being, I believe, according to recent surveys, only about 3,500 feet above the level of the sea; while, as before remarked, the other road passes over an elevation of 8,000 feet.

The beauty and capacity of the harbor of Puget Sound, the western terminus of the Northern Pacific Railroad, are too well known to need any further praise. It is the finest harbor in the world.

The road from this magnificent haven to water communication with the Atlantic, at the head of Lake Superior, will be only 1,600 or 1,700 miles long; where the chain of rivers, lakes and canals, from the Gulf of St. Lawrence and New York, meets the road at Superior Bay, 1,500 miles from the Atlantic tide-water, and nearly half way across the continent at its widest point. Such an advantage as this unparalleled water-course affords for freight and traffic for seven months in the year, no other route across the continent can ever possess. It is not my purpose to furnish accurate tables of comparative distances to show the saving in favor of the Northern Pacific over other routes from China to Europe, or from the Pacific to Chicago, New York, Boston or Portland. These are facts to be determined and settled by scientific men, by engineers, and by exact computation of curves, angles, and distances. But that is shorter it is evident, and that it is to become the great future highway of travel and trade between Europe and Eastern Asia, no man can doubt who properly appreciates the advantages of short lines and of direct water communication between London and Paris and the Northern Pacific Railroad at Superior Bay.

Independent of these great advantages for securing international commerce and travel, the Northern Pacific route is far superior to all others on account of the climate, productiveness of the soil, timber, minerals, the rivers, streams, and the water facilities afforded on the line. It is of some of these I design more particularly to speak, especially of the Northern Michigan and Mackinaw route.

It has long since been demonstrated by experience that wheat, barley, oats, hops, and all cereal productions, the potato, cabbage, beet, turnip, and other like vegetables, as well as apples, pears, plums, and some other fruits, grow more abundantly and are of better quality in the highest latitudes where they can mature, than in more southerly regions. For the production of most of these articles, especially the great staple of wheat, the vast basin or region of country lying between latitudes 44 and 54 north and longitudes 15 and 35 west from Washington, and known as the Saskatchewan, the Red River of the North and Upper Missouri river valleys, is unequalled by any other same number of acres upon the surface of the earth.

This region alone, (through the center of which the road is to pass,) when fully developed, is capable of supplying the world with grain, and of sustaining fifty millions of souls, a number much greater than the present population of the United States.

Of the timber, minerals, and soil upon the western slope, from the summit of the dividing mountain range to Puget Sound, in passing, I will only say that the testimony of all who have traveled over the route, and are familiar with the country bordering the same, is that its richness in these respects is only excelled by the grazing and wheat growing qualities of the eastern slope, from the mountains to Lake Superior.

Having thus far considered the subject from a general point of view, and having briefly referred to the country and its productive capabilities from Lake Superior to Puget Sound, I come now to consider the question as to the route the road should take from the head of Lake Superior to the Atlantic sea-board. From the head of Lake Superior to Montreal river, the dividing line between Michigan and Wisconsin, it is about 95 miles, and from that point to Mackinaw about 275 miles—making the distance from Superior to Mackinaw 370 miles, from Mackinaw to Saginaw 175 miles, from Saginaw to Detroit 96 miles, from Saginaw river to St. Clair river, on a line to Lewiston or Niagara, 80 miles; making the entire distance from Superior to Saginaw 545 miles, to Detroit 641, and to St. Clair river 625 miles. These distances, though not accurately determined by measurement, are believed to be substantially correct.

A road over this line, to connect at Lewiston with the one contemplated from Portland, must make a saving of distance, from Superior to that city, of about 200 miles over the route from Superior *via* Chicago. I am aware that this saving of distance will be diminished by a branch from the Northern Pacific, at some point considerably west of Superior, to Chicago; but Superior being the present eastern terminus fixed in the charter of the Northern Pacific for that road, and that being the great water outlet to the ocean, in summer, and the Mackinaw route offering the only competing railroad line for the shipment of the products of this vast northern region, it is of the greatest importance to the people of the same that this road should be built, even if the distance were no less than by Chicago.

The amount of local business that would be secured to such a road in Northern Michigan and Wisconsin, between Superior and the Saginaw river, seems to be but little understood or appreciated by the country, and I now desire to call the attention of this Convention and the railroad men of the East to the facts.

In the year 1835, or 1836, the Government of the United States (to settle the boundary dispute between Ohio and Michigan) ceded the Territory now known as the Upper Peninsula to the then Territory of Michigan. At this time, and for many years thereafter, very little was known of this important and valuable region, except that it was covered with timber and that copper had been discovered on Isle Royale and in some other localities, but as to the amount, or extent of its existence, no one knew. Nor is the value of its vast deposits of iron and copper known or appreciated even to this day.

As far back as 1636 copper was discovered by the Jesuit missionaries upon Lake Superior, and an attempt was made, in 1770, by an English company, to mine it. A vessel of 40 tons was built by this company upon Lake Superior, above the Sault, to transport men, machinery and supplies. Mining operations were commenced and prosecuted about two years near Ontonagon, but proving unprofitable, were abandoned in 1774.

The first scientific exploration of the mineral lands of Lake Superior was made by Dr. Douglass Houghton, Michigan State Geologist, between the years 1840 and 1845. He made annual reports of his discoveries, but as he perished, while prosecuting his labors, in 1845, by drowning, he never published his final report. The publication of his annual statements brought many to the district as early as 1843, and from that period to the present, copper mining has been pursued at many points on Lake Superior, and with varied successes, profits and losses.

Previous to this time, the entire freighting business of Lake Superior was carried on by one or two small schooners, which had to be built above the Sault. These rapids, in St. Mary's river, between Lakes Huron and Superior, are about one mile in length, and effectually prevented vessels from passing, by water, from the lower to the upper lake. In consequence of this, tram-ways were constructed around the falls, and several steamers hauled over the portage.

This slow and expensive process was rendered necessary, as no other means existed for obtaining vessels upon Lake Superior until the completion of the Sault Canal in 1855. That magnificent public work was built by the State of Michigan, through the aid of a grant of 750,000 acres of land made by the Government to the State, in 1853, for that purpose. At the period of the completion of the canal, and the free and uninterrupted passage of the largest vessels to and from Lake Superior, an immense impulse was given to the production and shipment of copper and iron. Before that time the expense of exportation made the production of these valuable minerals from their inexhaustible mines comparatively small. But, in the next eight years, as the following table will show, there was shipped from Lake Superior over $40,000,000 worth of copper alone:

	TONS.	POUNDS.
1855	7,642	
1855 to 1857	11,312	
1858	3,500	
1859	4,200	
1860	6,000	
1861	7,400	
1862	9,062	
1863	8,548	
1864	8,472	
1865	10,790	1,156
1866	10,375	1,684
1867	11,735	552
Total	90,037	1,396

This, as before remarked, brought in the market over $40,000,000; the product of 1867 bringing $4,700,000. During the present year (1869) there has been shipped from the Marquette Iron District about 600,000 tons of iron ore and pig metal, valued at $4,200,000.* The interest of copper has been greatly depressed, for the last few years, by the increased cost of production and for the want of as equal a protection as has been afforded to other similar interests. Under the protection now secured, it is hoped and believed that its production will increase until it shall reach an amount equal to the home demand of fifteen or sixteen thousand tons per annum.

* For the actual product of the Marquette Iron District, in 1869, see letter from Hiram A. Burt, Esq., in Appendix to this pamphlet.

With the increased facilities which the Northern Pacific Railroad will afford for a winter shipment of these articles, and for taking in supplies of all kinds for their production, there can be no reasonable doubt of such a result, nor that the annual production of pig iron and iron ore will be increased to 800,000 instead of 600,000 tons, as at present. When this shall be done, the annual exports of iron and copper will be twelve or thirteen million dollars.

About one-half of the iron is brought from the mountains to Marquette over the Marquette Railroad, and shipped from that point, the other half being taken to Escanaba, at the head of Green Bay, over the Escanaba Railroad, and shipped from there. The copper is principally shipped from Ontonagon, Houghton, Hancock, Eagle River, Eagle Harbor, Copper Harbor, Lac La Belle and Torch Lake. Marquette, Negaunee, Houghton, Hancock and Rockland are populous towns. Marquette and Negaunee, in the iron district, and Houghton and Hancock, in the copper, containing about 4,000 inhabitants each, and for schools, churches, hotels, stores and dwellings, will compare favorably with other towns of the same size. The whole population of the Upper Peninsula is believed to be not far from 60,000.

The number of vessels which passed through the Sault Canal in 1864 was 1,414, their united tonnage being 571,183. These vessels, forming regular lines between Buffalo, Cleveland, Detroit, Milwaukee and Chicago, were about one-quarter large class steamers, the balance being sailing vessels. The immense stores of all descriptions freighted through the canal to supply the mines and these inhabitants are purchased from the merchants, mechanics and traders of Pittsburg, Cleveland, Buffalo, Detroit, Chicago and Milwaukee, and are paid for in copper, iron ore, pig iron, lath, shingles, etc. Not only do the lake cities have the benefit of the profits on these supplies, but the copper and iron received in exchange gives employment to thousands of their mechanics, artisans, and laborers in their smelting works, foundries and machine shops, and from which other large direct and indirect profits are derived. In fact, it is doubtful whether there is any other source from which they receive so great a benefit.

The Upper Peninsula of Michigan can never become, to any great extent, an agricultural region, that portion bordering on the lake being too rocky and the soil too poor for many kinds of crops. In some portions of it, however, farming may hereafter be profitably pursued. Grass, cabbage, turnips, beets, and several other vegetables, grow to perfection. Wheat, oats and barley also ripen there, while potatoes of the best quality grow in great abundance. In September last, I saw some very fine apples, pears and plums growing on the farms of Messrs. Sayles, Mercer and Cash, at Ontonagon, and the vegetables spoken of in their gardens, as well as on the land of Mr. Hoyt, at Rockland. None of these, however, yet grow in sufficient quantities for home consumption except potatoes, which are beginning to be exported. There is, however, another source of wealth in the forests of white pine timber growing there; this, in consequence of its rapid exhaustion in all other sections of the country, is soon to become second in value to the mineral deposits.

So important has the Government of the United States considered the present and future products and commerce of Lake Superior to the country, that it has, from time to time, made large grants of land and money to deepen its rivers. and to build its ship canals, wagon and railroads. The principal grants for these objects have been as follows: For the Sault Canal, 750,000 acres; for the Portage Lake Ship Canal, 400,000 acres; for the Lac La Belle Canal, 100,000 acres; for the Marquette and Ontonagon Railroad, about 640,000 acres; for the Escanaba and Negaunee Railroad, about 320,000 acres; for the Chicago and Northwestern Railroad about 326,000 acres, and for the Fort Howard and Copper Harbor wagon road about 320,000 acres—making a total of appropriations of 2,856,000 acres of public lands. Of these, only about 1,600,000 acres have been actually acquired by the companies; 400,000 more for the Portage Lake and Lake Superior Ship Canal Company doubtless soon will be. The remaining 856,000 acres, for the west half of the Marquette and Ontonagon Railroad, from Lake Michigaumee to Ontonagon, for the Northwestern Railroad, from Menominee to Escanaba, and for the military wagon road from the Wisconsin State line to Fort Wilkins, at Copper Harbor. will soon revert to the Government, unless the grants are renewed. The money appropriated and recommended has been: For a breakwater at Marquette, $250,000; for Ontonagon harbor, $200,000, and for Eagle Harbor, $87,000—making $537,000. Besides this, large sums have been appropriated for the St. Mary's river, for light-houses, surveys, etc.

It has been carefully estimated, from as reliable data as can be ascertained, that the annual production of white pine lumber in the State of Michigan, including timber, shingles, and other descriptions, is greater than that produced in all the other States combined, and that there is still standing in its forests a larger amount, of a superior quality, than in all of the other territory east of the Rocky Mountains. This quality of pine, once growing in New York, Maine, and other eastern sections, has nearly all disappeared, leaving Michigan, Wisconsin and Minnesota to supply the present and rapidly increasing demand.

This almost indispensable and valuable timber once removed, is never to be replaced, at least to any extent; for if allowed to grow again, which is not the case, it would require from 50 to 80 years to become as fully grown as at present.

Fifteen years ago the amount of lumber manufactured in the State was, as compared with the present, very small. I am informed by Mr. Curtis Emerson, one of the earliest lumbermen here, that the only cargo of lumber shipped from the Saginaw Valley in 1846, twenty-three years ago, was on the schooner "Julia Smith," and consisted of 61,000 feet.

The following table will show the amount cut in the Valley in 1863, and the increase from that year to 1868, inclusive:

	FEET.
1863	133,500,000
1864	215,000,000
1865	250,639,340
1866	349,767,884
1867	423,963,190
1868	457,396,225
Total	1,830,266,639

It has been ascertained that in the year 1868 there was cut, in the entire State, 1,600 000,000 feet, Saginaw Valley and the Bay Shore producing about one-third of the whole amount. Making what I suppose to be a low estimate, that the *annual* production in the whole State, in the last eighteen years, has been 400,000,000 feet, the entire product within that period has been 7,200,000,000 feet. Placing the average yield at 3,750 feet to the acre, and at 300,000 feet to the 80-acre lot, we find the enormous number of 1,920,000 acres from which the pine has been removed in this State.

Estimating the entire amount yet standing in Northern Michigan, including the Upper Peninsula at double the amount, say 4,000,000 acres, the future yield will be 15,000,000,000 feet, and at the present price of $15 per M , will be worth in market $225,-000,000. The timber, shingles, etc., will bring at least $75,-000,000 more, making the pine in the forests of Northern Michigan produce the vast sum of $300,000,000. This 4,000,000 acres of standing pine, at present rates of exhaustion, will all be cleared in 12 or 14 years. But as heretofore the pine has been mostly cut on streams accessible for running logs, and as much of the remaining timber, both in the Lower and the Upper Peninsula, is found in sections too distant from the water-courses to make them available for that purpose, the supply from such localities must depend upon future railroad and other facilities for transportation to market. Until such means are afforded, the annual yield must soon be diminished, and this may prolong the entire exhaustion of white pine in the State for a period of 18 or 20 years; but beyond that, it is hardly possible.

Should the Northern Pacific Railroad be continued to Mackinac it will be about 2,000 miles long, and should its bonds be guaranteed by the Government for $30,000 per mile, (near the probable cost,) the entire guarantee would be $60,000,000, or one-fifth only of the value of the product of the pine forests of Northern Michigan.

The value of other timber can hardly be over-estimated, it being certain that no State in the Union has a larrger amount of white oak, maple, beech, ash, and other qualities of hard and soft timber, than Michigan. Of the article of oak staves there was manufactured and shipped from the Saginaw Valley alone in the year 1868:

6,846,000, which, at $60 per M., brought	$ 410,760
The hard wood timber of all kinds was valued at	300,000
Shingles, 104,104,500, at $4 per M.,	416,418
Lath,	123,000
Of the articles of headings, hoops, staves, etc., other than for salt barrels used here, there was shipped, in value,	50,000
Of pine lumber proper, 457,396,225 feet, at $15 per M.,	6,860,943
Making the total product of lumber, of all descriptions, shipped in 1868 from Saginaw Valley	$8,461,121

The manufacture of salt was commenced in the Valley in 1860, and the product has been as follows:

YEAR.	BARRELS.	YEAR.	BARRELS.
1860	4,000	1865	477,200
1861	125,000	1866	407 997
1862	243,000.	1867,	474,721
1863	466,356	1868	555,690
1864	529,073.		

Total number of barrels,		3,283,037
The product of 1868 amounting to	$1,111,380	
By adding the value of lumber,	8,461,121	
It makes the total product of the Valley in the articles of lumber and salt alone,		$9,572,501

In the whole State, including the above, there was manufactured, in 1868, 1,600,000,000 feet, at $15 per M., amounting to:

Lumber,	$24,000,000
Staves,	1,648,040
Timber,	1,000,000
Shingles,	1,249,254
Lath,	492,000
Hoops, heading, soft wood, staves, etc.,	150,000
Total	$28,534,294

OF FISH.

	BARRELS.
During the year 1868 there were caught at Thunder Bay	7,000
At Harrisville,	5,500
At the Sauble,	30,000
In the Saginaw river and at the head of the Bay,	4,000
Total,	46,500
And valued at	$325,500

Assuming this to be one-half the yield on the lake shores of the State, the value of the fish caught in the whole State in 1868 was $651,000.

PLASTER.

The extent and quality of the plaster beds at Alabaster, on the Saginaw Bay, are unsurpassed in this or any other State The exportation from there in 1868 was:

	TONS.
Of rock plaster,	25,000
Of ground do. for land,	2,000
Total	27,000
	BARRELS.
Of ground plaster,	10,000
Of ground do., calcined,	5,000
Total	15,000

[redacted]

I do not know the amount or value of the production of the Grand Rapids beds for this year, but the above exports from Alabaster brought in market $144,000.

From the foregoing statistics, which, when not taken from annual returns, are based upon careful estimates and comparisons, the gross value of the products of Northern Michigan, for the year 1868, inclusive of its agricultural, mechanical and several minor interests, was, from the forests, mines, quarries and fisheries alone, as follows:

Copper and iron from Lake Superior,	$ 8,900,000
Lumber, timber, shingles, lath, staves, etc.,	28,534,294
Salt,	1,111 380
Fish,	651,000
Plaster,	144,000
Total,	$39,340,594

These figures are certainly not exaggerated, but much lower than other statements made by competent men, who have spent more time and labor in consulting the facts, but whose calculations, in some respects are believed to be too high.

A Northern Pacific Railroad on the Mackinaw route would reach the extensive quarries of marble which have been discovered (and supposed to be of the best quality) at the head of Menominee river, thus affording an outlet for the same.

It would also run through that splendid pine region lying between the Escanaba and Montreal rivers, a large portion of which is not reached either by the Escanaba, Menominee, Montreal or Ontonagon rivers, or their tributaries.

These forests, as well as those in the Lower Peninsula, away from the streams and on the line of the road, must furnish thousands of millions of lumber for transportation by rail.

The richest and most extensive deposits of iron, in what is called the Marquette District, are known to be on the line of this road, between the head waters of the Menominee and Lake Michigaumee. These vast and inexhaustible stores are sufficient to produce a million tons per annum for a century to come, and all of which must be brought to market by rail.

Every worker of iron knows that the charcoal pig of Lake Superior is unequalled in quality, and that its production heretofore has been entirely inadequate to the demand. This is in consequence of the spareness of hard wood, and the soft and poor quality growing in the vicinity of the mines, for charcoal, and for supplying the furnaces with fuel. On the recent visit of Mr. S. P. Ely to this city, (the Superintendent of the Marquerte Railroad), knowing him to be a gentleman of much intelligence, and largely interested in the production and shipment of iron ore, I inquired of him how it was that they shipped the great bulk of iron in the ore, instead of first converting it into pig; to save the profit and prevent the loss of freight to Cleveland on the dross? His reply was as before stated, that it was for the want of hard wood, lime, &c., of which the lower Peninsula furnishes so vast an amount of the best quality. He added, "should a railroad be constructed across the Straits to the iron mines, furnaces and smelting works would be put in operation all along the line from Mackinaw to Saginaw, where the necessary materials for their successful operation are found without limitation as to quality or quantity."

This unquestionable result would not only furnish such a road with an unlimited amount of business, but would convert the forrests of hard wood into money instead of leaving them waste as at present; would clear the soil for the plow, rear villages, and people the line with thousands of industrious and thriving inhabitants. Not only so, but a Northern Pacific Railroad over the line suggested would soon be tapped by branches from the towns and mines on the copper range, and would thus secure the Spring, Fall and Winter travel and the freight of copper and supplies to and from the mines.

The forty million annual product of Northern Michigan would be increased to sixty millions, as most of the localities from which this product is realized would be reached by this road. If one-fifth of the passenger and freighting business was secured to the road, it would more than pay the entire cost of running from Superior to Saginaw River.

I was recently informed by an intelligent farmer, who cultivates a thousand acres about sixty miles west of St. Paul, in Minnesota, that the State produced twenty million bushels of wheat in 1868, and after much travel through that portion adapted to its growth, he gave it as his opinion, that not one acre in fifty had been cultivated. If, however, one.twentieth was used to produce the twenty million bushels, when all is sown with wheat the yield will be four hundred million of bushels

The Valleys of the Upper Missouri and Saskatchewan are as fine for the growth of wheat as Minnesota and embrace a territory five times as large. When they are peopled by agriculturist, as they soon will be, if railroad facilities are afforded, 2,000 million bushels of wheat can be produced, and all of which will seek an outlet by water, down the lakes, and by all the railroads that can be secured and reached.

This amount of wheat alone, as compared with the celebrated granaries of Egypt and the Black Sea, is as an ocean to a drop.

The Legislature of the State of Michigan, at its last session, passed an act ceding the Sault St. Mary Canal to the general Government, with a view of having it deepened to fourteen feet and made free to commerce. When this is done, and the Portage Canal, cutting off Keweenaw Point, is finished (thus materially shortening the distance, by water, from the Sault to the head of Lake Superior), and when the St. Paul and Superior, the Hudson, St. Croix and Superior and the Northern Pacific Railroads are completed to some point at the head of the lake, where a large city will be built, with its immense elevators and wharehouses to receive and discharge the freight, fifteen thousand vessels, instead of fifteen hundred, as at present, will annually pass through the Sault Canal.

The Mackinaw, and all other lines of Railroads from Chicago and other points connecting with the Northern Pacific, will require double tracks to do the business. This, too, without counting upon international or local business secured from the mineral regions west of the mountains.

If this line had been established, and, by reason of competition with other routes, had reduced the freight to the sea-board only five cents per bushel on wheat, the farmers of Minnesota would have saved thereby one million dollors on the crop of 1868.

The Trans-continental Railroad Convention recently held at Oswego, N. Y., appreciating the great necessity for the shortest and most direct Railroad connection with the Northern Pacific at the head of Lake Superior, after much able discussion and consideration, unamously adopted the following resolutions:

Resolved, by this Convention, That in forming a connection with the Northern Pacific Railroad at the head of Lake Superior, the shortest and most practical route should be adopted.

Resolved, That while in the opinion of this Convention, connections ought to and will be made *via* Chicago, the continuation of the N. P. R. R. to the Straits of Mackinaw, so as to connect with the contemplated road from Portland to the Detroit and St. Clair Rivers, ought also to be secured.

I am aware that the project of crossing the Sault River and passing through Canada to Montreal, and thence over to Vermont roads to Portland, has long been entertained by some of the officers of the N. P. R. R. Company, who are interested in said roads.

And I am also aware of the fact that this project becoming known, once defeated a subsidy to the N. P. Road by Congress.

This, while Canada remains a foreign country, ought not to be thought of. But that being the natural and by far the shortest route to the ocean, it can only in the nature of future events be delayed till Canada and the British possessions in North America become part of the United States. When that day comes, and come it soon must, the increased products of the North-West will create a necessity for that and all the other outlets to the East that can be secured, without injury to this or any other roads.

Finally, the only practical question to be considered is, *How shall this Convention aid in securing the speedy completion of the Northern Pacific Railroad by way of the Straits of Mackinaw?* I answer, by favoring in every possible, honorable way, Government aid to both such a Northern Pacific and a Southern Pacific Road. Such aid, either extended as security for the payment of the principal, or for the interest on the bonds of the roads, will secure their speedy completion. The Government can be amply secured against loss, by holding a mortgage on the lands and on all the property belonging to the corporations.

The liberal proposition will unite the strength of the North and South in Congress, and overcome any opposition the present Pacific Railroad might bring to bear against it. As a financial question in view of the National debt, if the roads are built, it is believed that the increase of population, of wealth, of commerce, and of taxable property, will make the nation richer instead of poorer, at the end of ten years, even if it should finally loose eighty or one hundred millions by becoming security.

This was the opinion expressed and maintained in a very able paper, prepared and read at the Oswego Convention, by Edwin F. Johnson, Esq., chief engineer of the Northern Pacific Railroad, and his views in this respect are concurred in by many prudent statesmen and thoughtful citizens. This eminent engineer also gives it as his opinion from practical knowledge, that there will be no serious difficulties met with in crossing the Straits of Mackinaw.

While living, Hon. Thaddeus Stevens was an ardent friend of the Northern Pacific Railroad by this route. He frequently told me when in conversation on the subject, that some day, not far in the future, one of the largest inland cities of the continent would be at the head of Lake Superior. In this opinion, I doubt not, he was correct.

At the conclusion of the reading of the foregoing paper, on motion of Mr. Hubbell, of Lake Superior, the thanks of the Convention were voted to Mr. Driggs, for his very able address.

Hon. D. H. Jerome, of Saginaw City, read the

REPORT OF THE COMMITTEE ON CREDENTIALS.

Showing the following named gentlemen to be entitled to seats as Delegates in this Convention: From

Ann Arbor—R. S. Smith.

Bay City—Jas Birney, H. M. Fitzhugh, C. C. Fitzhugh, D. H. Fitzhugh, Luther Westover, Wm. Westover, J. J. McCormick, J. McDermot, Geo. Campbell, J. J. Campbell.

Buena Vista—C. W. Kimball, Michael Leidlein, S. C Munson.

Chesaning—F. P. Kenyon, R. H. Nason, R. W. Andrus, J. C. Goodale, R. P. Mason, Geo. W. Chapman.

Corunna—James Cummins, Hugh McCurdy, S. R. Kelsey, J. N. Ingersoll, E. F. Wade, E. C. Moore, D. Bush, C. S. Converse, J. B. Wheeler, Wm. Oaks, John L. Simonson.

Chicago, Ill.—J. H. Pearson

Detroit—Wm H. Craig, Luther Beecher, Wm. Phelps, James P. Mansfield, E. Reidy, John Burt, R. H. Anderson, H. H. Emmons, J. D. Standish, F. W. Swift, Wm. A. Moore, Geo. S. Frost, Morgan Johnson, A E. Bissell, M. M. Fisher.

East Saginaw—Jas. L. Ketcham, John F. Driggs, Geo. W. Peck, J. Fisher, W. J Bartow, Joe A. Hollon, C. K. Robinson, Wm. L. Webber, D. W. C. Gage, J. K. Rose, L. S. Lenheim, J. F. Brown, A. B. Wood, C. H. Gage, M. B. Hess, T Edsall, David Rust, C. Ortman, M. Jeffers, E. T. Judd, J. J. Wheeler, T. E. Doughty, L. W. Tisdale.

Flint—A. Thayer, J. W. Begole, R. W. Jenny, A. L. Aldrich, F. H. Rankin, A. B. Witherbee.

Grand Rapids—Wm. A. Howard.

Houghton—J. A. Hubbell, Ransom Sheldon.

Holly—J. B. Simonson.

Jackson—Eugene Pringle, Horace S. Ismon.

Lapeer—E. J. White.

Lansing—J. W. Longyear, Cyrus Hewett, A. N. Hart, A. E. Cowles, Alfred Bixby, D. L. Case.

Marquette—Hirm A. Burt.

Midland—John Larkin, James W. Cochrane, John Haynes, C. E. Ball, James S. Eastman, George A. Janes.

Mason—O. M. Barnes.

Monroe—A. J. Sawyer, D. A. Noble, Edwin Willitts, T. R. Little, J. Wall, G. W. Strong, H. J. Redfield, C. Ives, J. R. Rauch, J. E. Ilginpitz, A. Darber, A. G. Bates, Charles Toll, E. G. Morton, T. Doyle, J. McLoure, M. D. Hamilton.

Northville—J. S. Lapham.

Oswego, N. Y.—A. S. Page, B. Doolittle, George Goble, R. Lippincott, C. Doolittle, Jno. A. Barry.

Owosso—B. O. Williams, D. Gould, A L. Williams.

Plymouth—C. H. Bennett.

Port Huron—O. D. Conger, J W. Sanborn, Edgar White, Cyrus Miles, John Miller, W. L. Bancroft, D. B. Harrington.

Portsmouth—Albert Miller.

Saginaw City—A. F. R. Braley, A. W. Wright, John Moore, D. H. Jerome, N. Barnard, W. S. Green, Daniel Hardin, C. T. Brenner, Thomas Merrill, Benton Hanchett, E. Briggs, James Haye, J. F. Adams, Ezra Rust, J. H. Sutherland.

Salem, State of Oregon—David Miller.

St. Clair—John E. Kilton, M. H. Miles, A. J. Chapman, Wm. Grace, Eugene Smith, B. W. Jenks.

St. Johns—Hon. Randolph Strickland.

Tuscola—J. H. Richardson, D. G. Slafter, W. H. Harrison, J. A. McPherson.

Toledo, Ohio.—A. E. McCumber, H. A. Boyd.

Vassar—B. W. Huston, Dr. Wm. Johnson, T. North, F. Bournes, Alexander Trotter, S. A. Lane.

Watrousville—H. P. Atwood.

Wenona—H. W. Sage, T. F. Sheppard, Jas. B. Ten Eyck, H. H. Wheeler, Frank Fitzhugh.

On motion, the report of the Committee was accepted and adopted.

Hon Wm. Phelps, of Detroit, presented, as the

REPORT OF THE COMMITTEE ON PERMANENT ORGANIZATION,

The names of the following persons as officers for the permanent organization of the Convention:

For President—Hon. John Moore, of Saginaw.

Vice Presidents—
John McLaren, of Monroe.
Josiah W Begole, of Flint.
John Burt, of Detroit.
Ransom Sheldon, of Houghton.
Horace S. Ismon, of Jackson.
B. O. Williams, of Owosso.
Marcus H. Miles, of St. Clair.
Luther Westover, of Bay City.
B. W. Huston, of Vassar.

Secretaries—
William Phelps, of Detroit.
F. H. Rankin, of Flint.
Michael Jeffers, of East Saginaw.

The report of the Committee was adopted; and the gentlemen named took their places on the platform.

On being introduced to the Convention by Mr. Driggs, Hon. John Moore, in assuming the Chair, made the following remarks:

Gentlemen of the Convention.—I am somewhat accustomed to the uncertainty that attends the rendition of a verdict by a petit jury. It has been said that before its rendition what such verdict will be is as uncertain as where lightning will strike. If I had received a slight shock of lightning, I should not have been more astonished than I am in being called upon to preside over the deliberations of this convention.

I appreciate the honor of being called upon to preside over a convention like this, of gentlemen assembled from all portions of our State and adjoining States, to deliberate upon a question of the importance of that submitted to the consideration of this convention. You have assembled irrespective of locality and irrespective of party predilections, to counsel together with regard to a great National enterprise.

The large attendance here to-day, and the interest already manifested in these proceedings, argue well for the success of the contemplated enterprise. The character of the gentlemen who have here assembled, show that the questions to be considered have engaged the attention of business men; that they are questions, not of an hour or a day, but that they are the developements of a great necessity, growing out of the unexampled growth of the business interests of our State and Nation. The paper, or report, which has been read to the convention by the Hon. John F. Driggs, giving as it does in detail a full statement of the great interests to be developed and promoted by the construction of the Northern Pacific Railroad, and the great benefits that will result to this State and the Nation at large, by such a road, have brought to your mind clearly the magnitude of the undertaking and the necessity of this great National enterprise.

There is no necessity that I should trespass upon your time by any further reference to the subject that has brought you together. It is one I have no doubt that has received your consideration. It could not have been otherwise. The importance of the enterprise is well understood. Any assistance that I can give you in your deliberations, or that I can hereafter render to promote the building of the road through our State, will be a pleasure on my part. With this feeling I accept the position you have assigned me, and thank you for the honor of serving you upon this platform.

Hon. George W. Peck moved that a Committee of seven members be appointed by the Chair, to report Resolutions, and an Order of Business, for the consideration of the Convention; which motion prevailed.

The President appointed as such

COMMITTEE ON RESOLUTIONS, ETC.

Geo. W. Peck and John F. Driggs, of East Saginaw; H. M. Fitzhugh, of Bay City; W. H. Craig, of Detroit; Artemas Thayer, of Flint; H. A. Burt, of Marquette; and Eugene Pringle of Jackson.

The Convention then adjourned until seven oclock P. M.

EVENING SESSION.

The Convention re-assembled pursuant to adjournment, and was called to order by the President.

Hon. Geo. W. Peck, for the Committee on Resolutions and Order of Business, announced that the Committee would not be prepared to report before to-morrow morning.

At the request of a Delegate, the Call under which this Convention had assembled, was read by one of the Secretaries.

An interesting discussion of the objects of the Convention then took place, and was continued to a late hour. It was participated in by Hon. Wm. A. Howard, of Grand Rapids; H. H. Emmons, Esq., of Detroit; Hon. John F. Driggs, of East Saginaw; Hon. Randolph Strickland, of St. Johns; and John A. Barry, Esq., of Oswego, N. Y.

At the close of Mr. Barry's remarks, on motion, the Convention adjourned until 10 o'clock to-morrow morning.

SECOND DAY.

MORNING SESSION.

The Convention was called to order at 10 o'clock, A. M. on Wednesday, November 24th.

Mr. John Burt, of Detroit, being called upon, addressed the Convention in favor of the route proposed, representing to the Convention the great facilities and advantages of the extension through the Upper Peninsula of Michigan, from information of the nature of the territory, derived from his personal explorations of many years, as a surveyor.

The President: There has just been placed in my hands a letter from Hon. Austin Blair, M. C., which the Secretary will read.

LETTER FROM EX-GOVERNOR BLAIR.

Jackson, Nov. 23, 1869.

Hon. Jas. L. Ketcham, John F. Driggs and other Committee, &c.

Gentlemen: I am wholly unable to attend the Convention at East Saginaw to-day, to which you were so kind as to invite me. My regret on this account is the greater that during the past summer I have had occasion to visit Puget Sound, and to become some acquainted with the needs and capabilities of that important region of our country.

The building of one railroad to the Pacific Ocean has served to prove how much we need two more. And especially to us of the northern portion of our country has it proved the necessity of the road to Puget Sound. To the great body of our people this is still to a great extent an unknown region, possessed of a cold and rigorous climate and valuable for little except lumber and timber, far away on the road to Alaska and eternal snow.

Nothing could be further from the truth. The climate of the country lying around this magnificent Sound is milder and more genial than that of our older States. The soil is rich and fertile, producing abundantly all the crops that belong to this zone. There are nearly fifteen hundred miles of coast all easily accessible to shipping, lying around this Sound and the bays that form a part of it.

Beyond all doubt here is to be the seat of a vast commerce and the homes of a numerous population. As a gateway to China and the Indies it is superior to the Bay of San Francisco.

There is no longer a doubt that the Northern route is more practicable at all seasons of the year than the Central. Besides it is *our route* above all others and cannot fail to commend itself to the good judgment of our people.

I should be glad to have been with you at this important convention, and to have joined in its discussions, but since that may not be, I shall hope still to aid you efficiently in other fields of your enterprise.

I sympathize entirely with your objects and will promote them in all proper ways.

Very Respectfully, Your Obedient Servant,

AUSTIN BLAIR.

MR. PECK: Before proceeding to discharge my duty as Chairman of the Committee on Resolutions, permit me to read to the Convention a letter from Mr. Johnson, the Engineer of the Northern Pacific Railroad.

LETTER FROM E. F. JOHNSON, C. E.

MIDDLETON, CT., Nov. 15th, 1869.

TO HON. JAS. L. KETCHAM, JOHN F. DRIGGS, and others, Committee, etc., East Saginaw, Michigan.

GENTLEMEN: It will not be in my power to comply with your invitation to be present at the convention at your place on the 23d inst., or to express in writing, at any length, my views upon the interesting topics to be discussed by the convention These, so far as the Northern Pacific road is concerned, were very fully stated to the convention lately held at Oswego, and will appear in their proceedings. For my general views as to the need of additional railway lines east of the Mississippi, and reasons therefor, I refer to my published report of last year, made to the Ontario Shore Railroad Company.

The Northern Pacific road, when built, is to have its eastern terminus, as provided in its charter, at some point on Lake Superior, within the States of Wisconsin or Minnesota. This point has not yet been determined. The Northern Pacific Company, so far as I know, do not contemplate asking Congress for any extension of their line towards the east beyond their present chartered limits. But if the country demands such extension, they will not, I presume, oppose it, but will carry out, in good faith, whatever is required of them by the Government.

Upon the subject of a line of railway from near the western extremity of Lake Superior to a connection with lines leading to Port Huron or Detroit, by the Straits of Mackinaw, I can say that, from my knowledge of the country south of Lake Superior, derived from observation and otherwise, the ground is in general favorable, and free from obstructions of a serious character, and will nowhere probably require a gradient exceeding 40 feet to the mile.

Passing from this region to Michigan proper, the principal impediment is the Straits of Mackinaw, which is three and one-half miles from Point La Barbe, on one side, to Point McGalpin upon the other, with an average depth of water from 65 to 70 feet, and a maximum depth of 216 feet, sufficient to deter from any attempt at bridging.

The passage of this strait by railway trains can, however, be very economically and expeditiously made, and, I think, at all seasons, by means of boats constructed for the purpose.

As to Michigan proper, or Lower Michigan, the face of the country is known to be very favorable for the construction of a road, with easy gradients. It is nowhere marked by any very strong features of a character to require extraordinary expense in the building of a road.

So far, therefore, as regards the line as a whole, the important question is whether the business and resources of the country through which it will pass are such as to justify its construction.

My own opinion is, that viewed as a local road only, it will eventually be well sustained. That portion of it lying west of Lake Michigan will have large mineral resources. The ores of iron now so extensively mined near Lake Superior are situated at Neguanee and vicinity, near the line of the railway to Marquette, on Lake Superior, on one side, and to Escanaba, on the little Bay de Noquet, on Lake Michigan, on the other.

From these mines are now shipped to southern and eastern markets over 300,000 tons of iron ore annually. To the south and west of this locality, near the divide of the Lake Superior and Mississippi waters, and extending west as far as the western extremity of the lake, there are other deposits of iron ore equally rich and extensive.

These latter, together with the copper, which is scattered throughout the same region will come within the influence of the proposed road and be developed by it. There are also in the same region large and valuable tracts of pine lands, the lumber of which will find its way to market over the road. The soil of this region in Wisconsin and especially in Upper Michigan, is in general, sandy and light, but there are considerable portions well adapted to cultivation which will be fully improved to supply the wants of the mining and lumbering population that will gather within its limits.

The line from Mackinaw, south, passes centrally through lower Michigan in a position favorable for commanding the business of that portion of the State. This portion is being rapidly settled. In addition to the farm products of the soil, it has extensive groves of pine and other timber and embraces within its limits a large part of the coal basin and the Salt Springs of Central Michigan.

The above remarks apply mainly to the local business of the proposed road. If an enterprise of this character will stand this test, its success may be deemed certain, for all lines of railway, wherever situated, invariably attract largely to their support the business of the more remote sections of the country.

Passing, as the proposed line does for much of its extent, through an undeveloped region, it will require the aid which the Government has, in similar cases, bestowed—a grant of the Government lands in alternate sections within a limited distance on either side of it. The aid thus given, when judiciously conferred upon lines that are really meritorious, enriches both the Government and the country. This is shown by the stimulus it has imparted to the building of railways in the newer sections of the country, and the rapid advance in those sections in population and in taxable wealth.

With best wishes that the action of your convention may result beneficially to Michigan and the country at large,

I am, gentlemen, very respectfully yours,

EDWIN F. JOHNSON,
Civil Engineer.

MR. RUST: I would remark to the Convention that Mr. Johnson's letter was written some time ago, and the past season shows an aggregate shipment of over 700,000 tons of iron ore.

MR. PECK: Mr. President—As Chairman of the Committee on Resolutions, I beg leave to report that the Committee deem it inexpedient to report any Rules, Regulations, or Order of Business.

They instruct me to submit Resolutions, which I will now proceed to read:

REPORT OF COMMITTEE ON RESOLUTIONS.

Resolved, That Railroads through the Northern Peninsula of the State of Michigan to the Straits of Mackinaw, and thence by direct connection with the East and North, are demanded by the interest, not only of Michigan, but of the whole country, and that such roads developing mining, lumbering, manufacturing, commercial, and agricultural interests of the first importance, will necessarily have the effect to transmute five hundred miles of wilderness into one of the most productive and richest sections of the Union.

Resolved, That such roads are necessary in justice to Michigan as a bond of Union between its two Peninsulas, and to continue by a practical and the shortest route towards the Atlantic, that future great continental highway, of which the Northern Pacific Road will form a part, and that in the judgment of this Convention it is the imparative duty of Congress to extend the land grants now held by the Jackson, Lansing & Saginaw, and the Grand Rapids and Indiana Railroad Companies, to the Straits of Mackinaw, relieving these Companies of the requirement to terminate upon Little Traverse Bay, and also to make such liberal grants of land, money or credit, as shall secure the speedy construction of the road from the present Eastern terminus of the Northern Pacific Road to the Straits of Mackinaw

Resolved, That the Convention deems it the true policy of the country to extend to such projects, as those of Railroads from Mackinaw to Puget Sound, and upon the line of the proposed Southern Pacific Road, such aid as will secure their speedy construction, and that we may cite the prosperity of such States and Districts as have been liberal in aiding great public works, as a perpetual argument to show the safety and wisdom of such a policy.

Resolved, That there be appointed by this Convention a committee of seven persons whose duty it shall be, by petitions to Congress, by explorations and by such continued public agitation of the subject, as may be necessary, to use all ligitimate efforts to secure the necessary legislation and investments of capital, and that it shall be the duty of such committee to appeal to the people and corporations more immediately interested for the means to defray the expense of such efforts.

Resolved, That while the said roads of Michigan and the Northwest, as pioneers in the developement of our own and neighboring State, (as far as their system is completed) are entitled to our highest consideration, and demand the warmest encouragement and sopport, from this Convention, and the people we represent; we do more especially and earnestly call upon Congress, to extend Government aid, in such efficient form as it may deem best, to secure the construction of the Northern Pacific Railroad to the Straits of Mackinaw, as the most important of all our Railroad interests.

Resolved, That the Committee appointed by this Convention, under the foregoing resolutions, be and they are hereby instructed, to institute inquiry into the condition, prospects, and purposes of the Northern Pacific Railroad Company, as at present organized, and if it shall satisfactorily appear, that the grant made to that Company, for the construction of its road, is likely to lapse or become forfeited; or such construction to be unreasonably delayed,—they are hereby instructed to bring the subject before Congress, and to urge its immediate action to compel or secure, the immediate commencement of the work in good faith; and its completion at the earliest practicable period.

Resolved, That the Secretaries of this Convention furnish to the President of the United States, and to the members of his Cabinet, certified copies of these resolutions, and printed copies of the proceedings of this Convention; and when Congress meets, furnish the President of the Senate, the Speaker of the House of Representatives, the Chairmen of the several Rail Road Committees; and of the Committees of Ways and Means, and Finance, respectively, with like copies of these resolutions and proceedings.

The report of the Committee on Resolutions was accepted, and a motion made and carried, that the Convention proceed to consider the resolutions separately and *seriatim*.

The first resolution being then under consideration, Mr. Luther Beecher, of Detroit, addressed the Convention at some length, and offered the following preamble and resolutions:

MR. BEECHER'S RESOLUTIONS.

WHEREAS, Large amounts of Government Land, in the States of Michigan and Wisconsin have for many years been in market, unsold and unpopulated to the great damage and detriment of large portions of both States; and

WHEREAS, Said Lands cannot be sold or occupied, even at the lowest graduation price of one shilling per acre, until water or Railway transit lines shall be constructed so as to pass through or near them; and

WHEREAS, These results cannot be otherwise so song as *this* state of things remains unchanged, by coming possibility prove advantageous to the nation; and

WHEREAS, The early extension of the Great Northern Pacific Railroad from the head of Lake Superior, Eastward, through the States of Wisconsin and Michigan is of vastly more importance to both States and of the whole Nation than any possible values of said lands; Therefore it is

Resolved, That it is for the National interest to surrender all of the unsold lands in each of said States to their respective Governors *in trust* the whole proceeds thereof to be expended in construction of such Trunk lines of Transit and such branches as shall best accomplish the ten fold purpose of selling and populating their land and aid in *construction of Railway and Water Transit lines*, demanded by the vast growth of the great North Western States.

Resolved, That said lands ought to be appraised and graded in value and *all* of them subject to sale for Cash or "Land Scrip", Freight, Transit, Bonds, Stock, or other evidences of value used in construction of said Trunk lines or Branches, and that the "Legislation" necessary for this purpose should be *prompt and of the most liberal character*.

Resolved, That as an offset for such surrender, of what is of no practical value to the General Government, but may be made of much value to the said States and of the Nation—in the manner proposed.—Said States can well afford to surrender, and forever relinquish, any and all unsettled war claims for bounties or other aid furnished by counties, town, cities or individuals—supposed to amount to about 3 or $4,000,000 *in each State*, while the lands, say 9,000,000 of acres in Wisconsin and 10.000,000 of acres in Michigan would only amount to about $3,000,000 at graduation value.

On motion, the preamble and resolutions of Mr. Beecher were referred to the Committee on Resolutions.

Hon. Eugene Pringle, of Jackson, was called upon, and discussed at length the resolution under consideration and the objects of the Convention.

At the close of Mr. Pringle's speech, the Convention adjourned until two o'clock in the afternoon.

AFTERNOON SESSION.

At two o'clock P. M., pursuant to adjournment, the Convention was called to order by the President.

Hon. O. M. Barnes, of Mason, and Dr. H. C. Potter, of East Saginaw, were severally called for, and addressed the Convention, illustrating their remarks by reference to a large map, produced by Mr. Barnes, showing the location, etc., of the land grants made to the Michigan Railroads.

The question was then taken upon the first resolution of the Committee, and the same was adopted.

The second resolution being again read, Mr. J. A. Hubbell, of Houghton, made a speech in its favor, after which the resolution was adopted.

The third resolution was then read, and was supported in a speech of some length by Hon. George W. Peck. The resolution was then adopted.

The fourth resolution was then adopted.

On the fifth resolution being read, Mr. Emmons moved to amend the resolution by striking out the words "to secure the construc- "tion of the Northern Pacific Railroad to the Straits of Macki- "naw," and insert, in lieu thereof, the words "to extend the route "of the Northern Pacific Railroad to the Straits of Mackinaw," and advocated his proposed amendment in remarks at some extent.

The amendment was opposed by Messrs. Peck, Driggs and Pringle, and was withdrawn by the mover.

Mr. Emmons asked the indulgence of the Convention while he offered a personal explanation; and proceeded to make some remarks in reference to his address to the Convention on the previous evening, elucidating more clearly the position he occupied.

The fifth resolution was then adopted.

The sixth and seventh resolutions were severally adopted.

THE PRESIDENT: Gentlemen of the Convention, by the resolutions you have passed, you have provided for the appointment of a Committee of Seven. The resolution does not specify how that Committee shall be appointed. I suppose, under ordinary Parliamentary rules, the Chair should appoint that Committee, but I decline to exercise that power. I think the Convention will see the impropriety of the exercise of that power by myself.

All local jealousies and feeling and controversy have been kept out of this Convention; and there are points along the line of this road that have their own local interests, which it is proper for them to look after. I reside at one of those points; and it might be claimed that I exercised the power which I would thus exercise to favor the ends of my own particular locality. I therefore insist that the Convention adopt some means by which that Committee shall be appointed.

Mr. Emmons moved that it be referred to the Committee on resolutions.

Mr. Peck moved as a substitute, that the Chair appoint a Committee of Five, to recommend the names of persons to the Convention, for appointment as a Committee required under the resolution.

The substitute was adopted, and the Chair appointed as such Committee: C. K. Robinson, of East Saginaw; C. C. Fitzhugh, of Bay City; H. A. Rust, of Marquette; W. H. Craig, of Detroit and W. S. Green, of Saginaw.

While awaiting the report of this Committee, the following additional letters on the objects of the meeting, were read to the Convention:

LETTER FROM ALFRED MEAD.

ONTONAGON, MICH., Nov. 13, 1869.

HON. JOHN F. DRIGGS,

DEAR SIR:

Your circular letter calling a Railroad Convention at Saginaw, on the 23d inst., has just reached me, and I at once send you a few lines to wish you and your friends God speed. It will be almost impossible for any one to attend from our county, (for you know we have no means of getting out and in the country at this time of the year), much as I wish that some of us would, had it been earlier or later, I would cheerfully have come myself, for you know I feel that a Railroad connecting us with the outer world is our only hope of relief from our present depressed condition.

But as we cannot attend, we shall feel confident that you, dear sir, will represent us and our views and watch our interests. Say to the gentlemen of that Convention that the people of Ontonagon will do all in their power to aid in the object of the Convention. We want a Railroad and will extend our aid to the first parties that approach us with any reasonable plan that will guarantee us the object, whether from the East, Michigan or Chicago. Of our situation and resources I need say but little, with them you are acquainted, also the attempts of Railroad Cos. to connect heretofore with us, but allow me to call your attention to one or two points. You are aware that the grant to the Ontonagon & Marquette Railroad Co. has been declared forfeited by the Legislature, this grant is now open for other parties to take hold of, it is a grant from the U. S. of ten Sections per mile, but twenty miles of that road has been built, and we venture to say, whoever will fill the gap between here and the road at these Iron Mines will reap large rewards. By the act of the Legislature any corporation upon whom this grant is conferred, are required to build fifteen miles of this end of the road, commencing at Ontonagon Village. To aid in the construction of the fifteen miles, this county as particularly interested, and we will lend our aid. To any corporation that will do it we will confer a State Swamp Land grant of two Sections per mile, the scrip is worth now, I believe, about 80 or 85 cents on the dollar, and is easily convertible into cash. From the best information we can obtain, the two Sections of Scrip will build the Road bed ready for the Iron, in fact, parties here (of no experience in road building) have offered to take the first ten miles, and get it ready for the iron, for the two Sections of Swamp Land.

As to the location of Ontonagon I am not the only one that thinks that it would be a favorable point for the commencement of the Northern Pacific. But I claim that road can have no *terminus* till it is connected with the rail road system of Michigan, and its most natural route, as through the Upper Peninsula, whether it will make connections at the Sault St. Marie or Mackinaw, it is for the people of Michigan to decide. The indications are that Old and New England Capital will make a desperate effort to cross at the Sault St. Marie, and the route has undoubtedly great advantages, but we shall see to it that it crosses our own territory. Hoping that the Convention may be well attended and its deliberations result in an effort to give us road out of the country. I am Sir, Yours very respectfully,

ALFRED MEAD.

LETTER FROM HIRAM WALKER.

To JAMES L. KETCHAM, Mayor of East Saginaw, and others.

MESSRS: I am in the receipt of your circular inviting the friends of the Northern Pacific Railroad, and its connections now made and to be made to form the shortest and most feasible route from Puget Sound, on the Pacific, to Portland, in Maine, on the Atlantic, to attend a Railroad Convention in your city on the 23d inst., which I am unable to attend.

No one at all informed on the "advance of Empire West" doubts the importance of these connecting lines of railroads, connecting the Atlantic and Pacific States, and whoever lives twenty years will see all these lines in operation. Who can tell the vast commercial interests that will flow through these channels? The Erie canal made the city of New York the metropolis of America, and these Pacific Railroads will make these United States the business center of the world. This is no myth.

The first railroad in the United States, of any note, was the Albany and Schenectady, which was completed in August, 1832; and the first line of telegraph in May, 1840, between Baltimore and Washington. And now what a network of railroads and telegraph lines do these States present. Computation would almost fail to show their cost of construction, much less to show the advance in value of real estate caused by their construction. San Francisco is nearer Boston, in point of time, than was the Valley of the Hudson one hundred years ago. If so much has been done in thirty-seven years, when large sums have been lost in experimenting, what grand results may be looked for in 1900.

Xerxes threw golden fetters into the sea to restrain the billows, in his anguish of defeat; but American skill and genius, with indomitable perseverance, has tunneled mountains, spanned mighty rivers, and strode triumphantly over the snow-clad Alps of America with the iron horse, which needed no rest from Plymouth Rock to the "Golden Gate." If such things are done in the "dry tree," what may we expect from the "green tree?"

"The wealth of Indies" will flow through these channels, aside from pecuniary gain. They will prove powerful agents of civilization by extending a knowledge of our free institutions. It will aid the missionary in the cause of evangelizing the world, and promote the cause of peace among the nations of the earth. Europe can reach China and Japan quicker and cheaper than by any other route, These roads will cement the interests of the different sections of this Union into one mighty Empire, and give us prestige among the nations of the earth; so that rock-bound New England, with its busy wheels and spindles, the rich alluvial soil of the Valley of the Mississippi, and the golden Pacific slope, may have a common interest in keeping this Government intact through ages to come.

Another consideration which should exert a powerful influence in favor of constructing this "Trans-Continental Railroad" is the fact that the Valley of the Mississippi, which is to be the great granary of the world, needs a cheaper and more ample outlet to the sea-board for her surplus products, and these products are increasing faster than the avenues of transportation are. So long as the producers can get a remunerative price for their products, so long they will patronize the mechanical interests of New England; but when they are compelled to retain them, for want of a ready outlet of their surplus products, at a fair home valuation, there will be a hegira of the wheels and spindles to the Mississippi Valley, which, like the "ten lost tribes," will never return.

There are two routes proposed by which to reach Lake Ontario from Portland. One is to strike Lake Champlain through Northern New Hampshire and Vermont, near Rouse's Point; thence by Ogdensburg to Oswego or Syracuse, N. Y. Another is to cross Connecticut river near Woodstock, Vt., thence to Rutland, Vt., and Whitehall, N. Y., at the head of Lake Champlain, thence to

Saratoga, to escape the spurs of the Adirondacks, thence westerly, north of the Central, to the Valley of Salmon river, thence to Oswego. The remainder of the route to St. Paul was too well delineated at the late Convention at Oswego to need repeating here.

Wishing you harmony and important action in your Convention about to meet, and that agitation will not cease until permanent measures are adopted for the construction of this important line of intercommunication,

I remain, gentleman,

Very respectfully yours,

HIRAM WALKER.

P. S. Please send me an account of your Convention.

MEXICO, N. Y., Nov. 16, 1869.

LETTER FROM WM. A. THOMPSON.

QUEENS HOTEL, TORONTO, Nov. 22, 1869.

JOHN F. DRIGGS, Esq.,
East Saginaw.

MY DEAR SIR :—I should certainly have responded to the invitation, and attended your Convention, but that I am tied down here by duties of the greatest importance, and which will hold me here for a fortnight When we get the Niagara, Detroit & St. Clair River Railway fairly floated, you will find me ready to respond to you, every effort of making a Great Through Line to the Northern Pacific *via* Saginaw.

I am, my dear sir, yours very truly,

WM. A. THOMSON.

You notice we are changing the name of our Railway.

MR. ROBINSON, from the Committee on nominations, reported the names of the following gentlemen, to compose the Committee of Seven, to be appointed under the provisions of the fourth resolution, viz :

STANDING COMMITTEE OF SEVEN.

JOHN F. DRIGGS, of East Saginaw.
JOHN RUST, of Detroit.
ARTEMAS THAYER, of Flint.
H. M. FITZHUGH, of Bay City.
JOHN MOORE, of Saginaw City.
J. A. HUBBELL, of Houghton.
S. P. ELY, of Marquette.

On motion, the report was accepted, and the nominations of the Committee, as above, ratified.

Mr. Driggs regreted to be obliged positively to decline acting as Chairman of the Committee; and Mr. Moore also asked to be excused from serving as a member; but the Convention refused to make any change in the constitution of the Committee as reported.

On motion the members of the Committee were authorized to appoint substitutes, in case any of them should be unable to attend to the business assigned them.

On motion of Hon. George W. Peck, the Secretaries were instructed to have not less than one thousand copies of the proceedings of the Convention printed in pamphlet form, for distribution.

The thanks of the Convention were voted to the President, Vice-Presidents and Secretaries, for the faithful performance of their duties; to the Reporters of the Press, for their full and impartial reports of the proceedings; to Capt. Kirby and Messrs. E. G. Godard and W. H. Herbert for furnishing the large and complete map used by the Convention; and to Mr. W. J. Bartow, for the use of Irving Hall.

The President, in a few remarks, thanked the Convention for the very harmonious manner in which they had conducted their business, and wished them success in the great undertaking they had inaugurated.

On motion, the Convention adjourned *sine die.*

JOHN MOORE, *President.*

JOHN MCLAREN,
JOSIAH W. BEGOLE,
JOHN BURT,
RANSOM SHELDON,
HORACE S. ISMON,
B. O. WILLIAMS,
MARCUS H. MILES,
LUTHER WESTOVER,
B. W. HUSTON,
} *Vice Presidents.*

WILLIAM PHELPS,
F. H. RANKIN,
MICHAEL JEFFERS,
} *Secretaries.*

[redacted]

APPENDIX.

Actual Products of Iron Ore and Pig Metal in the Marquette District.

From the Saginaw Daily Enterprise, of December 1st, 1869.

East Saginaw, Nov. 30, 1869.

To the Editor of the Enterprise.

Sir:—I send you for publication a letter just received from Hiram A. Burt, Esq., Collector of Customs, at Marquette, Lake Superior. It will be seen that his statement of the yield of iron in the Lake Superior region during the present year, 1869, makes the same nearly one hundred thousand tons more than given in your table of figures in the paper read to the Convention. It was my purpose in the paper referred to (and which was prepared without much time) to be as accurate as possible, and where my figures were not made up from actual returns, to place them below rather than above what subsequent facts might prove to be correct. Before placing the product of iron at 600,000 tons. I inquired of Mr. S. P. Ely, of Marquette, as to the amount for 1869, and he informed me that while he did not yet know the exact amount, I would certainly be safe in placing it at 600,000 tons.

It now appears from Mr. Burt's statement, which is doubtless correct, that the yield is 696,030 tons, valued at $ 5,185,398, being in excess over the value as fixed by my figures, $ 935,398. I hardly think it possible that my calculations as to lumber and other products, will be found so much a varience with the facts, as they are based upon returns of former years. But I am confident, if not strictly correct, they will prove to be below instead of above the mark. The larger yield of iron, it will be seen, makes the product of Northern Michigan over $ 40,000,000.

J. F. DRIGGS.

LETTER FROM H. A. BURT.

HON. JOHN F. DRIGGS, East Saginaw.

DEAR SIR:—I notice, on reading the published report of your able paper, read before the late Railroad Convention in your city, that you so much under-estimate the present year's production of iron ore and pig metal in the Marquette Iron District, that I desire to give you the actual figures up to Nov. 15th.

The actual product of Iron Ore is	629 489 gross tons.		
Estimated product balance of year	30 000 "		
Total product Ore		657 487	gross tons.
Pig Metal to Nov. 15th,	35 543 gross tons.		
Estimated product balance of year	3 000 "	38 543	"
Total of Ore and Pig Metal		696 030	"
Value		$ 5 185 898	
Increased production over 1869		149 971	
" value "		$ 1 182 985	

With regard to future production, I know of blast furnaces enough now building to increase the demand for ore next year fully 150,000 tons, thus realizing, in 1870, your prophecy as to what must take place upon the construction of the Northean Pacific Railroad; 800,000 tons in 1870, and 1,000,000 in 1872, are my prediction of the requirement from the Lake Superior Iron Mines in those years. What would be the resnlt of the construction of the Northern Pacific Railroad is beyond calculation.

Yours, Respectfully,

HIRAM A. BURT.

Spotted something that isn't quite right?

Let us know and we'll fix it for future reprints.
And as a thank you for your diligence, we'll
send you a free Sothis book of your choice.

publisher@sothis.press

This page intentionally left blank

Except of course for that line above. And this one.
And whilst we have your attention, perhaps you'd
like to follow us on Instagram @sothis.press. Go on.

Made in the USA
Monee, IL
08 November 2026

68089561R00104